SpringerBriefs in Computer Science

Series Editors

Stan Zdonik
Peng Ning
Shashi Shekhar
Jonathan Katz
Xindong Wu
Lakhmi C. Jain
David Padua
Xuemin Shen
Borko Furht
V. S. Subrahmanian
Martial Hebert
Katsushi Ikeuchi
Bruno Siciliano

For further volumes:
http://www.springer.com/series/10028

Tao Li · Shigang Chen

Traffic Measurement
on the Internet

 Springer

Tao Li
University of Florida
Gainesville, FL
USA

Shigang Chen
University of Florida
Gainesville, FL
USA

ISSN 2191-5768   ISSN 2191-5776 (electronic)
ISBN 978-1-4614-4850-1 ISBN 978-1-4614-4851-8 (eBook)
DOI 10.1007/978-1-4614-4851-8
Springer New York Heidelberg Dordrecht London

Library of Congress Control Number: 2012943363

Printed on acid-free paper

Springer is part of Springer Science+Business Media (www.springer.com)

Acknowledgment

Research work included in the book was supported in part by the US National Science Foundation under grant CNS-1115548.

Contents

Chapter 1
Introduction

Abstract Traffic measurement provides critical real-world data for service providers and network administrators to perform capacity planning, accounting and billing, anomaly detection, and service provision. In many measurement functions, statistical methods play important roles in system designing, model building, formula deriving, and error analyzing. One of the greatests challenges in designing an online measurement function is to minimize the per-packet processing time in order to keep up with the line speed of the modern routers. To meet this challenge, one should minimize the number of memory accesses per packet and implement the measurement module in the on-die cache memory. Hence, it is critical to make the data structures of a measurement module as compact as possible. This book presents several novel online measurement methods that are compact and fast.

Keywords Internet traffic measurement · Compact and fast solutions

1.1 Online Network Functions

Modern high-speed routers forward packets from incoming ports to outgoing ports via switching fabric, bypassing main memory and CPU. New technologies are pushing line speeds beyond OC-768 (40 Gb/s) to reach 100 Gb/s or even tera bits per second [14]. The line cards in core routers must therefore forward packets at a rate exceeding 150 Mpps [35]; that leaves little time to process each packet. Parallel processing and pipeline are used to speed up packet switching to a few clock cycles per packet [15]. In order to keep up with such high throughput, online network functions for traffic measurement, packet scheduling, access control, and quality of service will also have to be implemented using on-chip cache memory and bypassing main memory and CPU almost entirely [22, 35, 45]. However, fitting these network functions in fast but small on-chip memory represents a major technical challenge today [15, 29].

T. Li and S. Chen, *Traffic Measurement on the Internet*,
SpringerBriefs in Computer Science, DOI: 10.1007/978-1-4614-4851-8_1,
© The Author(s) 2012

1

The commonly-used cache memory on network processor chips is SRAM, typically a few megabytes. Further increasing on-chip memory to more than 10 MB is technically feasible, but it comes with a much higher price tag and access time is longer. There is a huge incentive to keep on-chip memory small because smaller memory can be made faster and cheaper. Off-chip SRAM is larger. For example, QDR-III SRAM has 36 MB [27]. But it is slower to access. Hence, on-chip memory remains the first choice for online network functions that are designed to match the line speeds.

On-chip memory is limited in size. To make the matter even more challenging, it may have to be shared by security [18], measurement [22], routing [6], and performance [17] functions that are implemented on the same chip. When multiple network functions share the same memory, each of them can only use a fraction of the available space. Depending on their relative importance, some functions may be allocated tiny portions of the limited memory, whereas the amount of data they have to process and store can be extremely large in high-speed networks. The disparity in memory demand and supply requires us to implement online functions as compact as possible [36, 39]. Furthermore, when different functions share the same memory, they may have to take turns to access the memory, making memory access the performance bottleneck. Since most online functions require only simple computations that can be efficiently implemented in hardware, their throughput will be determined by the bottleneck in memory access. Hence, we must also minimize the number of memory accesses made by each function when it processes a packet. The challenge is that compactness (in terms of space requirement) and speed (in terms of memory accesses) are sometimes conflicting objectives.

1.2 Fundamental Primitives

The implementations of many online functions heavily rely on several fundamental building blocks for data processing and storage. This book studies three important fundamental online functions: per-flow size estimators, spread estimators, and origin-destination flow estimators.

Per-flow size estimators are used to measure per-flow information for high-speed links. The goal is to estimate the size of each flow (in terms of number of packets). A flow is identified by a label that can be a source address, a destination address, or any combination of addresses, ports, and other fields in the packet header. Measuring the sizes of individual flows has important applications. For example, if we use the addresses of the users as flow labels, per-flow size measurement provides the basis for usage-based billing and graceful service differentiation, where a user's service priority gracefully drops as he over-spends his resource quota. Studying per-flow data over consecutive measurement periods may help us discover network access patterns and, together with user profiling, reveal geographic/demographic traffic distributions among users. Such information will help Internet service providers and application developers to align network resource allocation with the majority's needs.

We define a *contact* as a source-destination pair, for which the source sends a packet to the destination. The source or destination can be an IP address, a port number, a combination of address/port together with other fields in the packet header, or even a file name or URL in the payload. The *spread of a source* is the number of *distinct destinations* contacted by the source during a measurement period. Similarly, we can define the *spread of a destination*, which is the number of *distinct sources* that have contacted the destination. Measuring spread values has many applications. Intrusion detection systems can use them to detect port scans [36], in which an external host attempts to establish too many connections to different internal hosts or different ports of the same host. They may be used to detect DDoS attacks when too many hosts send traffic to a receiver [28], i.e., the spread of a destination is abnormally high. They can be used to estimate the infection rate of a worm by monitoring how many addresses each infected host contacts over a period of time. A large server farm may use the spread values of its servers to find how popular the servers' content is, which provides guidance for resource allocation. An institutional gateway may monitor outbound traffic and identify external web servers that have large spread values. This information helps the local proxy learn the popularity of servers and determine the cache priority of web content.

Origin-destination (OD) flow estimators are used to measure *OD flow sizes*. Consider two routers r_1 and r_2. We define the set of packets that first pass r_1 and then pass r_2 or first pass r_2 and then pass r_1 as an *origin-destination (OD) flow* of the two routers. The cardinality of the packet set is called the *OD flow size*. The OD flow measurement is also an important topic in many network management applications [9, 13, 25, 31, 32]. For example, Internet service providers may use the OD-flow information between points of interest as a reference to align traffic distribution within the network. They may also study the OD-flow traffic pattern and identify anomalies that deviate significantly from the normal pattern. In the event of a persistent congestion, OD-flow data may help point out the source of the congestion.

One of the greatest challenges in designing an online measurement module is to minimize the per-packet processing time in order to keep up with the line speed of the modern routers. To meet this challenge, we should minimize the number of memory accesses per packet and implement the measurement module in the on-die SRAM, which is fast but expensive. Because many other functions may also run from SRAM, it is expected that the amount of high-speed memory allocated for the module will be small. Hence, it is critical to make the measurement module's data structure as compact as possible.

1.3 Per-Flow Size Estimation Through Randomized Counter Sharing

This book first presents a particularly challenging problem, the measurement of per-flow sizes for a high-speed link without using per-flow data structures [20]. It has been shown in [10] that maintaining per-flow counters cannot scale for high-speed links.

Even for efficient counter implementations [30, 33, 46], SRAM will only be able to hold a small fraction of per-flow state (including counters and indexing data structures such as pointers and flow identities for locating the counters). The *counter braids* avoid per-flow counters and achieve near-optimal memory efficiency [22, 23]. This method maps each flow to a certain number of arbitrary counters; they are all incremented by one for every packet of the flow. Many flows may be mapped to the same counter, which stores the sum of the flow sizes. Essentially, the counters represent linear equations, which can be solved for the flow sizes. Two levels of counters are used to reduce the memory overhead. The counter braids require slightly more than 4 bits per flow and are able to count the exact sizes of all flows. But it also has two limitations. First, it performs 6 or occasionally 12 memory accesses per packet. Second, when the memory allocated to a measurement function is far less than 4 bits per flow, the message passing decoding algorithm of counter braids cannot converge to any meaningful results. When the available memory is just 1∼2 bits per flow, the *exact measurement* of the flow sizes is no longer possible. We have to resort to *estimation methods*. The key is to efficiently utilize the limited space to improve the accuracy of the estimated flow sizes, and do so with the minimum number of memory accesses per packet.

We present a fast and compact per-flow size estimation function that achieves three main objectives: (i) It shares counters among flows to save space, and does not incur any space overhead for mapping flows to their counters. This distinguishes our work from [30, 33, 46]. (ii) It updates exactly one counter per packet, which is optimal. This separates our work from the counter braids that update three or more counters per packet. Updating each counter requires two memory accesses for read and then write. (iii) It provides estimation of the flow sizes, as well as the confidence intervals that characterize the accuracy, even when the available memory is too small such that other exact-counting methods including [22, 23] no longer work. We believe this is the first size estimator that achieves all these objectives. It complements the existing work by providing additional flexibility for the practitioners to choose when other methods cannot meet the speed and space requirements.

The design of our size estimator is based on a new data encoding/decoding scheme, called *randomized counter sharing*. It splits the size of each flow among a number of counters that are randomly selected from a counter pool. These counters form the *storage vector* of the flow. For each packet of a flow, we randomly select a counter from the flow's storage vector and increment the counter by one. Such a simple online operation can be implemented very efficiently. The storage vectors of different flows share counters uniformly at random; the size information of one flow in a counter is the noise to other flows that share the same counter. Fortunately, this noise can be quantitatively measured and removed through statistical methods, which allow us to estimate the size of a flow from the information in its storage vector. We present two estimation methods whose accuracies are statistically guaranteed. They work well even when the total number of counters in the pool is by far smaller than the total number of flows that share the counters. The experimental results based on real traffic traces demonstrate that the new methods can achieve good accuracy in a tight space. We also provide several methods to increase the range of flow sizes that the estimators can measure.

The randomized counter sharing scheme presented in this work for per-flow size measurement has applications beyond the networking field. It may be used in the data streaming applications to collect per-item information from a stream of data items.

1.4 Spreader Classification

It is very costly to measure the spread of each source (or destination) precisely. When a router measures the spread of a source, it has to remember the destinations that the source has contacted so far. Future packets from the source to the same destinations do not increase the spread value. The spread is increased only when a packet is sent to a new destination. The problem is that it takes too much memory to store all destination addresses that every source has contacted.

To solve this problem, various techniques such as sampling [38], probabilistic counting [16], Bloom filters [45], and bitmaps [3, 11, 39] are used to reduce memory overhead at the expense of measurement accuracy. The rationale is that absolutely precise measurement of spread values may not be necessary for most applications. It is often practically sufficient to estimate spread values with a certain level of accuracy. Moreover, many applications only require us to classify spreaders into categories, such as (1) *heavy spreaders*, i.e., sources (or destinations) whose spread values are large, and (2) *non-heavy spreaders*. This further lowers the accuracy requirement and allows additional room for memory saving. For example, in scan detection, we want to identify heavy spreaders (scanners) that have contacted a lot of destinations. In the previous server-farm example, we want to know the set of servers with large spread values. Even if we do not identify all such servers, it is very helpful in resource allocation if we can identify most of them.

This book addresses the *spreader classification* problem. *Single-objective spreader classification* is to identify the set of heavy spreaders. *Multi-objective spreader classification* places sources (or destinations) into more categories based on their spread values. We present an efficient spreader classification scheme based on a new storage method, called *dynamic bit sharing*, which stores contact information of all sources in a compact format. The level of compactness is so deep that the total number of available bits is less than one twentieth of the number of sources in some of our experiment cases: on average, just one bit is available for every twenty sources. Yet still we are able to make spreader classification with predictable accuracy. We employ a maximum likelihood estimation method to extract per-source information from the compact storage and determine the heavy spreaders. It ensures that false positive/negative ratios are bounded. Moreover, given an arbitrary set of false positive/negative bounds, we develop a systematic approach to determine the optimal system parameters, such that the amount of memory needed to satisfy the bounds is minimized. We carry out experiments based on a real traffic trace and demonstrate that, using these optimal parameters, we can reduce the memory consumption by three to twenty times when comparing with other existing work.

1.5 Origin-Destination Flow Measurement

When we solve the problem of *origin-destination (OD) flow measurement* [21], the goal is to design an efficient method to measure the number of packets that traverse between two routers during a measurement period. It generally consists of two phases: One for online packet information storage and the other for offline OD-flow size computation. In the first phase, routers record information about arrival packets. In the second phase, each router reports its stored information to a centralized server, which performs the measurement of each OD flow based on the information sent from the origin/destination router pair.

Measurement efficiency and accuracy are two main technical challenges. In terms of efficiency, we want to minimize the per-packet processing overhead to accommodate future routers that forward packets at extremely high rates. More specifically, the function should minimize the computational complexity and the number of memory accesses for each packet.

Accuracy is another important design goal. In high-speed networks, we have to deal with a very large volume of packets. And it is unrealistic to store all packet-level information in order to achieve 100 % accuracy. To solve this problem, some past research [40–42] uses data such as link load, network routing, and configuration data to indirectly measure the OD flows. Cao et al. [2] propose a quasi-likelihood approach based on a continuous variant of the Flajolet-Martin sketches [12]. However, none of them is able to achieve both efficiency and accuracy at the same time.

To meet these challenges, we present a novel OD flow measurement method, which uses a compact bitmap data structure for packet information storage. At the end of a measurement period, bitmaps from all routers are sent to a centralized server, which examines the bitmaps of each origin/destination router pair and uses a statistical inference approach to estimate the OD flow size. The new method has three elegant properties. First, its processing overhead is small and constant, only one hash operation and one memory access per packet. Second, it is able to achieve excellent measurement results, which will be demonstrated by both simulations and experiments. Finally, its data storage is very compact. The memory allocation is less than 1 bit for each packet on average.

Traffic measurement is an important subject of Internet technologies. In the broad context of computer networks, there are many other topics such as QoS and maxmin routing [5, 7, 8, 24, 26, 34, 37], P2P networks [19, 43, 44], distributed computing [1, 4], etc. Although we do not address these topics, they may interact with traffic measurement under certain scenarios where new research problems and applications may sprout.

1.6 Outline of the Book

The rest of the book is organized as follows: Chap. 2 presents a fast and compact per-flow size estimator based on *randomized counter sharing*. In this chapter, we provide of a novel data encoding/decoding scheme, which mixes per-flow information randomly in a tight SRAM space for compactness. Chapter 3 presents an efficient spread estimation scheme based on *dynamic bit sharing*, which optimally combines probabilistic sampling, bit-sharing storage, and maximum likelihood estimation. Chapter 4 gives a novel method for OD flow measurement which employs the bitmap data structure for packet information storage and uses statistical inference approach to compute the measurement results.

References

1. Basu, A., Buch, V., Vogels, W., von Eicken, T.: U-Net: a user-level network interface for parallel and distributed computing. In: Proceedings of the ACM SOSP, pp. 40–53 (1995)
2. Cao, J., Chen, A., Bu, T.: A quasi-likelihood approach for accurate traffic matrix estimation in a high speed network. In: Proceedings of IEEE INFOCOM (2008)
3. Cao, J., Jin, Y., Chen, A., Bu, T., Zhang, Z.: Identifying high cardinality internet hosts. In: Proceedings of the IEEE INFOCOM (2009)
4. Chen, S., Deng, Y., Attie, P., Sun, W.: Optimal deadlock detection in distributed systems based on locally constructed wait-for graphs. In: Proceedings of the 16th International Conference on Distributed, Computing Systems, pp. 613–619 (1996)
5. Chen, S., Fang, Y., Xia, Y.: Lexicographic maxmin fairness for data collection in wireless sensor networks. IEEE Trans. Mob. Comput. 6(7), 762–776 (2007)
6. Chen, S., Nahrstedt, K.: Maxmin fair routing in connection-oriented networks. In: Proceedings of the Euro-Parallel and Distributed Systems, pp. 163–168 (1998)
7. Chen, S., Shavitt, Y.: SoMR: a scalable distributed QoS multicast routing protocol. J. Parallel Distrib. Syst. 68(2), 137–149 (2008)
8. Chen, S., Song, M., Sahni, S.: Two techniques for fast computation of constrained shortest paths. IEEE/ACM Trans. Netw. 16(1), 105–115 (2008)
9. Erramilli, V., Crovella, M., Taft, N.: An independent-connection model for traffic matrices. In: Proceedings of the Internet Measurement Conference (IMC) (2006)
10. Estan, C., Varghese, G.: New directions in traffic measurement and accounting. In: Proceedings of the ACM SIGCOMM (2002)
11. Estan, C., Varghese, G., Fish, M.: Bitmap algorithms for counting active flows on high-speed links. IEEE/ACM Trans. Netw. (TON) 14(5), 925–937 (2006)
12. Flajolet, G.: Probabilistic counting. In: Proceedings of the Symposium on Fundations of Computer Science (FOCS) (1983)
13. Fortz, B., Thorup, M.: Optimizing OSPF/IS-IS weights in a changing world. IEEE JSAC Special Issue on Advances in Fundamentals of Network Management (2002)
14. Gardner, W.D.: Researchers transmit optical data at 16.4 Tbps. InformationWeek (2008)
15. Hermsmeyer, C., Song, H., Gemelli, R., Bunse, S.: Towards 100G packet processing: challenges and technologies. Bell Labs Tech. J. 14(2), 57–80 (2009)
16. Hwang, K., Vander-Zanden, B., Taylor, H.: A linear-time probabilistic counting algorithm for database applications. ACM Trans. Database Syst. 15(2), 208–229 (1990)
17. Jian, Y., Chen, S.: Can CSMA/CA networks be made fair? In: Proceedings of the 14th ACM International Conference on Mobile Computing and Networking, pp. 235–246 (2008)

18. Jung, J., Paxson, V., Berger, A., Balakrishnan, H.: Fast portscan detection using sequential hypothesis testing. In: Proceedings of the IEEE Symposium on Security and Privacy (2004)
19. Kamvar, S., Schlosser, M., Garcia-Molina, H.: The eigentrust algorithm for reputation management in P2P networks. In: Proceedings of the World Wide Web Conference (2003)
20. Li, T., Chen, S., Ling, Y.: Fast and compact per-flow traffic measurement through randomized counter sharing. In: Proceedings of the IEEE INFOCOM (2011)
21. Li, T., Chen, S., Qiao, Y.: Origin-Destination flow measurement in high-speed networks. In: Proceedings of the IEEE INFOCOM, Mini-Conference (2011)
22. Lu, Y., Montanari, A., Prabhakar, B., Dharmapurikar, S., Kabbani, A. Counter braids: a novel counter architecture for per-flow measurement. In: Proceedings of the ACM SIGMETRICS (2008)
23. Lu, Y., Prabhakar, B.: Robust counting via counter braids: an error-resilient network measurement architecture. In: Proceedings of the IEEE INFOCOM (2009)
24. Lui, K., Nahrstedt, K., Chen, S.: Hierarchical QoS routing in delay-bandwidth sensitive networks. In: Proceedings of the 25th Annual IEEE Conference on Local, Computer Networks, pp. 579–588 (2000)
25. Medina, A., Taft, N., Salamatian, K., Bhattacharyya, S., Diot, C.: Traffic matrix estimation: existing techniques and new directions. In: Proceedings of the ACM SIGCOMM (2002)
26. Nahrstedt, K., Chen, S.: Coexistence of QoS and best-effort flows-routing and scheduling. In: Proceedings of the 10th IEEE Tyrrhenian International Workshop on Digital Communications: Multimedia, Communications (1998)
27. Pearson, M.: QDRTM-III: Next generation SRAM for networking. http://www.qdrconsortium. org/presentation/QDR-III-SRAM.pdf (2009)
28. Plonka, D.: FlowScan: a network traffic flow reporting and visualization tool. In: Proceedings of the USENIX LISA (2000)
29. Qiao, Y., Li, T., Chen, S.: One memory access bloom filters and their generalization. In: Proceedings of the IEEE INFOCOM (2011)
30. Ramabhadran, S., Varghese, G.: Efficient implementation of a statistics counter architecture. In: Proceedings of the ACM SIGMETRICS (2003)
31. Ringberg, H., Soule, A., Rexford, J., Diot, C.: Sensitivity of PCA for traffic anomaly detection. In: Proceedings of the ACM SIGMETRICS (2007)
32. Roughan, M., Thorup, M., Zhang, Y.: Traffic engineering with estimated traffic matrices. In: Proceedings of the Internet Measurement Conference (IMC) (2003)
33. Shah, D., Iyer, S., Prabhakar, B., McKeown, N.: Maintaining statistics counters in router line cards. IEEE Micro. **22**(1), 76–81 (2002)
34. Shi, Y.: Hou, Y.T.: Theoretical results on base station movement problem for sensor network. In: Proceedings of the IEEE INFOCOM (2008)
35. Song, H., Hao, F., Kodialam, M., Lakshman, T.: IPv6 lookups using distributed and load balanced bloom filters for 100 Gbps core router line cards. In: Proceedings of the IEEE INFOCOM (2009)
36. Staniford, S., Hoagland, J., McAlerney, J.: Practical automated detection of stealthy portscans. J. Comput. Secur. **10**(1–2), 105–136 (2002)
37. Tang, Y., Chen, S., Ling, Y.: State aggregation of large network domains. Comput. Commun. **30**(4), 873–885 (2007)
38. Venkatataman, S., Song, D., Gibbons, P., Blum, A.: New streaming algorithms for fast detection of superspreaders. In: Proceedings of the NDSS (2005)
39. Yoon, M., Li, T., Chen, S., Peir, J.: Fit a spread estimator in small memory. In: Proceedings of the IEEE INFOCOM (2009)
40. Zhang, Y., Roughan, M., Duffield, N., Greenberg, A.: Fast accurate computation of large-scale ip traffic matrices from link loads. In: Proceedings of the ACM SIGMETRICS (2003)
41. Zhang, Y., Roughan, M., Lund, C., Donoho, D.: An information theoretic approach to traffic matrix estimation. In: Proceedings of the ACM SIGCOMM (2003)
42. Zhang, Y., Roughan, M., Lund, C., Donoho, D.: Estimating point-to-point and point-to-multipoint traffic matrices: an information-theoretic approach. IEEE/ACM Trans. Networking **13**(5), 947–960 (2005)

43. Zhang, Z., Chen, S., Ling, Y., Chow, R.: Capacity-aware multicast algorithms on heterogeneous overlay networks. IEEE Trans. Parallel Distrib. Syst. **17**(2), 135–147 (2006)
44. Zhang, Z., Chen, S., Yoon, M.: MARCH: A distributed incentive scheme for peer-to-peer networks. In: Proceedings of the IEEE INFOCOM, pp. 1091–1099 (2007)
45. Zhao, Q., Xu, J., Kumar, A.: Detection of super sources and destinations in high-speed networks: Algorithms, analysis and evaluation. IEEE J. Sel. Areas Commun. (JASC) **24**(10), 1840–1852 (2006)
46. Zhao, Q., Xu, J., Liu, Z.: Design of a novel statistics counter architecture with optimal space and time efficiency. In: Proceedings of the ACM Sigmetrics/Performance (2006)

Chapter 2
Per-Flow Size Estimators

Abstract This chapter discusses the measurement of per-flow sizes for high-speed links. It is a particularly difficult problem because of the need to process and store a huge amount of information, which makes it difficult for the measurement module to fit in the small but fast SRAM space (in order to operate at the line speed). We provide a novel measurement function that estimates the sizes of all flows. It delivers good performance in tight memory space where other approaches no longer work. The effectiveness of the online per-flow measurement approach is analyzed and confirmed through extensive experiments based on real network traffic traces. The rest of this chapter is organized as follows: Sect. 2.1 discusses the performance metrics. Section 2.2 gives an overview of the system design. Section 2.3 discusses the state of the art. Section 2.4 presents the online data encoding module. Sections 2.5–2.6 present two offline data decoding modules. Section 2.7 discusses the problem of setting counter length. Section 2.8 addresses the problem of collecting flow labels. Section 2.9 presents the experimental results. Section 2.10 extends the estimators for large flow sizes. Section 2.11 gives the summary.

Keywords Per-flow size estimator, Randomized counter sharing

2.1 Performance Metrics

We measure the number of packets in each flow during a measurement period, which ends every time after a certain number (e.g., 10 millions) of packets are processed. The design of per-flow measurement functions should consider the following three key performance metrics.

T. Li and S. Chen, *Traffic Measurement on the Internet*,
SpringerBriefs in Computer Science, DOI: 10.1007/978-1-4614-4851-8_2,
© The Author(s) 2012

2.1.1 Processing Time

The per-packet processing time of an online measurement function determines the maximum packet throughput that the function can operate at. It should be made as small as possible in order to keep up with the line speed. This is especially true when multiple routing, security, measurement, and resource management functions share SRAM and processing circuits.

The processing time is mainly determined by the number of memory accesses and the number of hash computations (which can be efficiently implemented in hardware [24]). The counter braids [21, 22] update k counters at the first level for each packet. When a counter at the first level overflows, it needs to update k additional counters at the second level. If $k = 3$, it requires at least 3 hashes and 6 memory accesses to read and then write back after counter increment. In the worse case, it requires 6 hashes and 12 memory accesses. The multi-resolution space-code Bloom filters [19] probabilistically select one or more of its 9 filters and set 3~6 bits in each of the selected ones. Each of those bits requires one memory access and one hash computation.

Our objective is to achieve a constant per-packet processing time of one hash computation and two memory accesses (for updating a single counter). This is the minimum processing time for any method that uses hash operations to identify counters for update.

2.1.2 Storage Overhead

The need to reduce the SRAM overhead has been discussed in Chap. 1. One may argue that because the amount of memory needed is related to the number of packets in a measurement period, we can reduce the memory requirement by shortening the measurement period. However, when the measurement period is smaller, more flows will span multiple periods and consequently the average flow size in each period will be smaller. When we measure the flow sizes, we also need to capture the flow labels [22], e.g., a tuple of source address/port and destination address/port to identify a TCP flow. The flow labels are too large to fit in SRAM. They have to be stored in DRAM. Therefore, in a measurement period, each flow incurs at least one DRAM access to store its flow label. If the average flow size is large enough, the overhead of this DRAM access will be amortized over many packets of a flow. However, if the average flow size is too small, the DRAM access will become the performance bottleneck that seriously limits the throughput of the measurement function. This means the measurement period should not be too small.

2.1.3 Estimation Accuracy

Let s be the size of a flow and $\hat{s}$ be the estimated size of the flow based on a measurement function. The estimation accuracy of the function can be specified by a confidence interval: the probability for s to be within $[\hat{s} \cdot (1 - \beta), \hat{s} \cdot (1 + \beta)]$ is at least a pre-specified value α, e.g., 95 %. A smaller value of β means that the estimated flow size is more accurate in a probabilistic sense.

There is a tradeoff between estimation accuracy and storage overhead. If the storage space and the processing time are unrestricted, we can accurately count each packet to achieve perfect accuracy. However, in practice, there will be constraints on both storage and processing speed, which make 100 % accurate measurement sometimes infeasible. In this case, one has to settle with imperfect results that can be produced with the available resources. Within the bounds of the limited resources, we must explore novel measurement methods to make the estimated flow sizes as accurate as possible.

2.2 System Design

2.2.1 Basic Idea

We use an example to illustrate the idea behind the new measurement approach. Suppose the amount of SRAM allocated to one of the measurement functions is 2 Mb (2×2^{20} bits), and each measurement period ends after 10 million packets, which translate into about 8 s for an OC-192 link (10+ Gbps) with an average packet size of 1,000 bytes. The types of flows that the online functions may measure include per-source flows, per-destination flows, per-source/destination flows, TCP flows, WWW flows (with destination port 80), etc. Without losing generality, suppose the specific function under consideration in this example measures the sizes of TCP flows.

Figure 2.1 shows the number of TCP flows that have a certain flow size in log scale, based on a real network trace captured by the main gateway of our campus. If we use 10 bits for each counter, there will be only 0.2 million counters. The number of concurrent flows in the trace for a typical measurement period is around 1 million. It is obvious that allocating per-flow state is not possible and each counter has to store the information of multiple flows. But if an "elephant" flow is mapped to a counter, that counter will overflow and lose information. On the other hand, if only a couple of "mouse" flows are mapped to a counter, the counter will be under-utilized, with most of its high-order bits left as zeros.

To solve the above problems, we not only store the information of multiple flows in each counter, but also store the information of each flow in a large number of counters, such that an "elephant" is broken into many "mice" that are stored at different counters. More specifically, we map each flow to a set of l randomly-selected counters and split the flow size into l roughly-equal shares, each of which

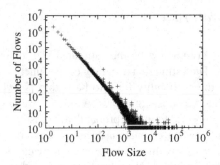

Fig. 2.1 Traffic distribution: each point shows the number (*y coordinate*) of flows that have a certain size (*x coordinate*)

is stored in one counter. The value of a counter is the sum of the shares from all flows that are mapped to the counter. Because flows share counters, they introduce noise to each other's measurement. The key to accurately estimate the size of a flow is to measure the noise introduced by other flows in the counters that the flow is mapped to.

Fortunately, this can be done if the flows are mapped to the counters uniformly at random. Any two flows will have the same probability of sharing counters, which means that each flow will have the same probability of introducing a certain amount of noise to any other flow. If the number of flows and the number of counters are very large, the combined noise introduced by all flows will be distributed across the counter space about uniformly. The statistically uniform distribution of the noise can be measured and removed. The above scheme of information storage and recovery is called *randomized counter sharing*.

We stress that this design philosophy of "splitting" each flow among a large number of counters is very different from "replicating" each flow in k counters as the counting Bloom filter [7] or counter braids [21] do—they add the size of each flow as a whole to k randomly selected counters. Most notably, the method increments one counter for each arrival packet, while the counting Bloom filter or counter braids increment k counters. We store the information of each flow in many counters (e.g., 50), while they store the information of each flow in a small number of counters.

2.2.2 Overall Design

The online traffic measurement function consists of two modules. The online data encoding module stores the information of arrival packets in an array of counters. For each packet, it performs one hash function to identify a counter and then updates the counter with two memory accesses, one for reading and the other for writing. At the end of each measurement period, the counter array is stored to the disk and then reset to zeros.

The offline data decoding module answers queries for flow sizes. It is performed by a designated offline computer. We present two methods for separating the information about the size of a flow from the noise in the counters. The first one is called the *counter sum estimation method* (CSM), which is very simple and easy to compute. The second one is called the *maximum likelihood estimation method* (MLM), which is more accurate but also more computationally intensive. The two complementary methods provide flexibility in designing a practical system, which may first use CSM for rough estimations and then apply MLM to the ones of interest.

2.3 State of the Art

A related thread of research is to collect statistical information of the flows [11, 18], or identify the largest flows and devote the available memory to measure their sizes while ignoring the smaller ones [9, 12, 15, 16]. For example, RATE [17] and ACCEL-RATE [14] measure per-flow rate by maintaining per-flow state, but they use a *two-run sampling* method to filter out small-rate flows so that only high-rate flows are measured.

Another thread of research is to maintain a large number of counters to track various networking information. One possible solution [8, 25] can be statistically update a counter according to the current counter size. This approach is suitable for the applications with loose measurement accuracy. In order to enhance the accuracy performance, Zhao et al. [29] propose a statistical method to make a DRAM-based solution practical. Since large DRAM is involved, this approach is able to achieve decent measurement accuracy.

Also related is the work [26] that measures the number of *distinct* destinations that each source has contacted. Per-flow counters cannot be used to solve this problem because they cannot remove duplicate packets. If a source sends 1,000 packets to a destination, the packets contribute only one contact, but will count as 1,000 when we measure the flow size. To remove duplicates, bitmaps (instead of counters) should be used [5, 13, 27, 28, 30]. From the technical point of view, this represents a separate line of research, which employs a different set of data structures and analytical tools. Attempt has also been made to use bitmaps for estimating the flow sizes, which is however far less efficient than counters, as our experiments will show.

2.4 Online Data Encoding

The flow size information is stored in an array C of m counters. The ith counter in the array is denoted as $C[i]$, $0 \leq i \leq m - 1$. The size of the counters should be set so that the chance of overflow is negligible; we will discuss this issue in details later. Each flow is mapped to l counters that are randomly selected from C through hash functions. These counters logically form a *storage vector* of the flow, denoted as C_f, where f is the label of the flow. The ith counter of the vector, denoted as $C_f[i], 0 \leq i \leq l - 1$, is selected from C as follows:

$$C_f[i] = C[H_i(f)], \tag{2.1}$$

where $H_i(...)$ is a hash function whose range is $[0, m)$. We want to stress that C_f is *not* a separate array for flow f. It is merely a logical construction from counters in C for the purpose of simplifying the presentation. In all the formulas, one should treat the notation $C_f[i]$ simply as $C[H_i(f)]$. The hash function $H_i, 0 \leq i \leq l - 1$, can be implemented from a master function $H(...)$ as follows: $H_i(f) = H(f|i)$ or $H_i(f) = H(f \oplus R[i])$, where '|' is the concatenation operator, '$\oplus$' is the XOR operator, and $R[i]$ is a constant whose bits differ randomly for different indices i.

All counters are initialized to zeros at the beginning of each measurement period. The operation of online data encoding is very simple: When the router receives a packet, it extracts the flow label f from the packet header, randomly selects a counter from C_f, and increases the counter by one. More specifically, the router randomly picks a number i between 0 and $l - 1$, computes the hash $H_i(f)$, and increases the counter $C[H_i(f)]$, which is physically in the array C, but logically the ith element in the vector C_f.

2.5 Offline Counter Sum Estimation

2.5.1 Estimation Method

At the end of a measurement period, the router stores the counter array C to a disk for long-term storage and offline data analysis.

Let n be the combined size of all flows, which is $\sum_{i=0}^{m-1} C[i]$.

Let s be the true size of a flow f during the measurement period. The estimated size, $\hat{s}$, based on the counter sum estimation method (CSM) is

$$\hat{s} = \sum_{i=0}^{l-1} C_f[i] - l\frac{n}{m}. \tag{2.2}$$

The first item is the sum of the counters in the storage vector of flow f. It can also be interpreted as the sum of the flow size s and the noise from other flows due to counter sharing. The second item captures the expected noise. Below we formally derive (2.2).

Consider an arbitrary counter in the storage vector of flow f. We treat the value of the counter as a random variable X. Let Y be the portion of X contributed by the packets of flow f, and Z be the portion of X contributed by the packets of other flows. Obviously, $X = Y + Z$.

Each of the s packets in flow f has a probability of $\frac{1}{l}$ to increase the value of the counter by one. Hence, Y follows a binomial distribution:

$$Y \sim Bino(s, \frac{1}{l}) \qquad (2.3)$$

Each packet of another flow f' has a probability of $\frac{1}{m}$ to increase the counter by one. That is because the probability for the counter to belong to the storage vector of flow f' is $\frac{l}{m}$, and if that happens, the counter has a probability of $\frac{1}{l}$ to be selected for increment. Assume there is a large number of flows, the size of each flow is negligible when comparing with the total size of all flows, and l is large such that each flow's size is randomly spread among many counters. We can approximately treat the packets independently. Hence, Z approximately follows a binomial distribution:

$$Z \sim Bino(n - s, \frac{1}{m}) \approx Bino(n, \frac{1}{m}), \text{ because } s \ll n. \qquad (2.4)$$

We must have

$$E(X) = E(Y + Z) = E(Y) + E(Z) = \frac{s}{l} + \frac{n}{m}. \qquad (2.5)$$

That is,

$$s = l \times E(X) - l\frac{n}{m}. \qquad (2.6)$$

From the observed counter values $C_f[i]$, $E(X)$ can be measured as $\frac{\sum_{i=0}^{l-1} C_f[i]}{l}$. We have the following estimation for s:

$$\hat{s} = \sum_{i=0}^{l-1} C_f[i] - l\frac{n}{m}. \qquad (2.7)$$

If a flow shares a counter with an "elephant" flow, its size estimation can be skewed. However, the experiments show that CSM works well in general because the number of "elephants" is typically small (as shown in Fig. 2.1) and thus their impact is also small, particularly when there are a very large number of counters and flows. Moreover, the next method based on maximum likelihood estimation can effectively reduce the impact of an outlier in a flow's storage vector that is caused by an "elephant" flow.

2.5.2 Estimation Accuracy

We derive the mean and variance of $\hat{s}$ as follows: Because $X = Y + Z$, we have

$$E(X^2) = E((Y+Z)^2) = E(Y^2) + 2E(YZ) + E(Z^2)$$
$$= E(Y^2) + 2E(Y)E(Z) + E(Z^2)$$
$$= \frac{s^2}{l^2} - \frac{s}{l^2} + \frac{s}{l} + 2 \cdot \frac{s}{l} \cdot \frac{n}{m} + \frac{n^2}{m^2} - \frac{n}{m^2} + \frac{n}{m}.$$

The following facts are used in the above mathematical process: $E(Y^2) = \frac{s^2}{l^2} - \frac{s}{l^2} + \frac{s}{l}$ because $Y \sim Bino(s, 1/l)$. $E(YZ) = E(Y)E(Z)$ since Y and Z are independent. $E(Z^2) = \frac{n^2}{m^2} - \frac{n}{m^2} + \frac{n}{m}$ because $Z \sim Bino(n, 1/m)$.

$$Var(X) = E(X^2) - (E(X))^2$$
$$= \frac{s}{l}(1 - \frac{1}{l}) + \frac{n}{m}(1 - \frac{1}{m}). \tag{2.8}$$

In (2.7), $C_f[i]$, $0 \le i \le l-1$, are independent samples of X. We can interpret $\hat{s}$ as a random variable in the sense that a different set of samples of X may result in a different value of $\hat{s}$. From (2.7), we have

$$E(\hat{s}) = l \times E(X) - l\frac{n}{m}$$
$$= l(\frac{s}{l} + \frac{n}{m}) - l\frac{n}{m} = s, \tag{2.9}$$

which means our estimation is unbiased. The variance of $\hat{s}$ can be written as

$$Var(\hat{s}) = l^2 \times Var(X) = l^2\left(\frac{s}{l}(1 - \frac{1}{l}) + \frac{n}{m}(1 - \frac{1}{m})\right)$$
$$= s(l - 1) + l^2\frac{n}{m}(1 - \frac{1}{m}). \tag{2.10}$$

2.5.3 Confidence Interval

We derive the confidence interval for the estimation as follows: The binomial distribution, $Z \sim Bino(n, 1/m)$, can be closely approximated as a Gaussian distribution, $Norm(\frac{n}{m}, \frac{n}{m}(1 - \frac{1}{m}))$, when n is large. Similarly, the binomial distribution, $Y \sim Bino(s, \frac{1}{l})$, can be approximated by $Norm(\frac{s}{l}, \frac{s}{l}(1 - \frac{1}{l}))$. Because the linear combination of two independent Gaussian random variables is also normally distributed [6], we have $X \sim Norm(\frac{s}{l} + \frac{n}{m}, \frac{s}{l}(1 - \frac{1}{l}) + \frac{n}{m}(1 - \frac{1}{m}))$. To simplify the presentation, let $\mu = \frac{s}{l} + \frac{n}{m}$ and $\Delta = \frac{s}{l}(1 - \frac{1}{l}) + \frac{n}{m}(1 - \frac{1}{m})$.

$$X \sim Norm(\mu, \Delta), \tag{2.11}$$

where the mean μ and the variance Δ agree with (2.5) and (2.8), respectively. Because $\hat{s}$ is a linear function of $C_f[i]$, $0 \le i \le l - 1$, which are independent samples of X, $\hat{s}$ must also approximately follow a Gaussian distribution. From (2.7) and (2.11), we have

$$\hat{s} \sim Norm(s, s(l-1) + l^2 \frac{n}{m}(1 - \frac{1}{m})).\tag{2.12}$$

Hence, the confidence interval is

$$\hat{s} \pm Z_\alpha \sqrt{s(l-1) + l^2 \frac{n}{m}(1 - \frac{1}{m})},\tag{2.13}$$

where α is the confidence level and Z_α is the α percentile for the standard Gaussian distribution. As an example, when $\alpha = 95\%$, $Z_\alpha = 1.96$.

2.6 Maximum Likelihood Estimation

In this section, we provide the second estimation method that is more accurate but also more computationally expensive.

2.6.1 Estimation Method

We know from the previous section that any counter in the storage vector of flow f can be represented by a random variable X, which is the sum of Y and Z, where $Y \sim Bino(s, \frac{1}{l})$ and $Z \sim Bino(n, 1/m)$. For any integer $z \in [0, n)$, the probability for the event $Z = z$ to occur can be computed as follows:

$$Pr\{Z = z\} = \binom{n}{z}(\frac{1}{m})^z(1 - \frac{1}{m})^{n-z}.$$

Because n and m are known, $Pr\{Z = z\}$ is a function of a single variable z and thus denoted as $P(z)$.

Based on the probability distribution of Y and Z, the probability for the observed value of a counter, $C_f[i]$, $\forall i \in [0, l)$, to occur is

$$Pr\{X = C_f[i]\}$$

$$= \sum_{z=0}^{C_f[i]} (Pr\{Z = z\} \cdot Pr\{Y = C_f[i] - z\})$$

$$= \sum_{z=0}^{C_f[i]} \binom{s}{C_f[i] - z} (\frac{1}{l})^{C_f[i]-z} (1 - \frac{1}{l})^{s-(C_f[i]-z)} P(z). \qquad (2.14)$$

Let $y = C_f[i] - z$ to simplify the formula. The probability for all observed values in the storage vector of flow f to occur is

$$L = \prod_{i=0}^{l-1} Pr\{X = C_f[i]\}$$

$$= \prod_{i=0}^{l-1} \left(\sum_{z=0}^{C_f[i]} \binom{s}{y} (\frac{1}{l})^y (1 - \frac{1}{l})^{s-y} P(z) \right). \qquad (2.15)$$

The maximum likelihood method (MLM) is to find an estimated size $\hat{s}$ of flow f that maximizes the above likelihood function. Namely, we want to find

$$\hat{s} = \arg \max_{s} \{L\}. \qquad (2.16)$$

To find $\hat{s}$, we first apply logarithm to turn the right side of the equation from product to summation.

$$\ln(L) = \sum_{i=0}^{l-1} \ln \left(\sum_{z=0}^{C_f[i]} \binom{s}{y} (\frac{1}{l})^y (1 - \frac{1}{l})^{s-y} P(z) \right). \qquad (2.17)$$

Because $\frac{d\binom{s}{y}}{ds} = \binom{s}{y}(\psi(s+1) - \psi(s+1-y))$, where $\psi(...)$ is the polygamma function [1], we have

$$\frac{d(\binom{s}{y}(1 - \frac{1}{l})^{s-y})}{ds} =$$

$$\binom{s}{y}(1 - \frac{1}{l})^{s-y} \left(\psi(s+1) - \psi(s+1-y) + \ln(1 - \frac{1}{l}) \right).$$

To simplify the presentation, we denote the right side of the above equation as $O(s)$. From (2.17), we can compute the first-order derivative of $\ln(L)$ as follows:

$$\frac{d \ln(L)}{ds} = \sum_{i=0}^{l-1} \frac{\sum_{z=0}^{C_f[i]} \left(O(s)(\frac{1}{l})^y P(z) \right)}{\sum_{z=0}^{C_f[i]} \binom{s}{y}(\frac{1}{l})^y (1 - \frac{1}{l})^{s-y} P(z)}. \tag{2.18}$$

Maximizing L is equivalent to maximizing $\ln(L)$. Hence, by setting the right side of (2.18) to zero, we can find the value for $\hat{s}$ through numerical methods. Because $\frac{d \ln(L)}{ds}$ is a monotone function of s, we can use the bisection search method to find the value $\hat{s}$ that makes $\frac{d \ln(L)}{ds}$ equal to zero.

2.6.2 Estimation Accuracy

We derive the estimation confidence interval as follows: The estimation formula is given in (2.16). According to the classical theory for MLM, when l is sufficiently large, the distribution of the flow-size estimation $\hat{s}$ can be approximated by

$$Norm(s, \frac{1}{\mathcal{I}(\hat{s})}), \tag{2.19}$$

where the *fisher information* $\mathcal{I}(\hat{s})$ [20] of L is defined as follows:

$$\mathcal{I}(\hat{s}) = -E\left(\frac{d^2 \ln(L)}{ds^2} \right). \tag{2.20}$$

In order to compute the second-order derivative, we begin from (2.11) and have the following:

$$Pr\{X = C_f[i]\} = \frac{1}{\sqrt{2\pi \Delta}} e^{-\frac{(C_f[i]-\mu)^2}{2\Delta}}$$

$$\ln(Pr\{X = C_f[i]\}) = -\ln(\sqrt{2\pi \Delta}) - \frac{(C_f[i] - \mu)^2}{2\Delta}, \tag{2.21}$$

where $0 \le i \le l - 1$. Performing the second-order differentiation, we have

$$\frac{d^2 \ln(Pr\{X = C_f[i]\})}{ds^2} = -\frac{\mu'}{l\Delta} + \frac{(\frac{1}{2}(1 - \frac{1}{l}) + \mu - C_f[i])\Delta'}{l\Delta^2}$$
$$+ \frac{1}{l\Delta^3}(1 - \frac{1}{l})\left((\mu - C_f[i])\mu'\Delta - (\mu - C_f[i])^2\Delta' \right), \tag{2.22}$$

where $\mu' = \frac{1}{l}$ and $\Delta' = \frac{1}{l}(1 - \frac{1}{l})$. Therefore,

$$E(\frac{d^2 \ln(Pr\{X = C_f[i]\})}{ds^2})$$

$$= -\frac{\mu'}{l\Delta} + \frac{\frac{1}{2}(1 - \frac{1}{l})\Delta'}{l\Delta^2} + \frac{1}{l\Delta^3}(1 - \frac{1}{l})E(\mu - C_f[i])^2\Delta'$$

$$= -\frac{1}{l^2\Delta} + \frac{3(1 - \frac{1}{l})^2}{2l^2\Delta^2}, \tag{2.23}$$

where we have used the following facts: $E(\mu - C_f[i]) = 0$ and $E(\mu - C_f[i])^2 = \Delta$. Because $L = \prod_{i=0}^{l-1} Pr\{X = C_f[i]\}$, we have

$$\mathcal{I}(\hat{s}) = -E\left(\frac{d^2 \ln(L)}{ds^2}\right) = \sum_{i=0}^{l-1} E(\frac{d^2 \ln(Pr\{X = C_f[i]\})}{ds^2})$$

$$= \frac{1}{l\Delta} - \frac{3(1 - \frac{1}{l})^2}{2l\Delta^2}. \tag{2.24}$$

From (2.19), the variance of $\hat{s}$ is

$$Var(\hat{s}) = \frac{1}{\mathcal{I}(\hat{s})} = \frac{2l\Delta^2}{2\Delta - 3(1 - \frac{1}{l})^2}. \tag{2.25}$$

Hence, the confidence interval is

$$\hat{s} \pm Z_\alpha \cdot \sqrt{\frac{2l\Delta^2}{2\Delta - 3(1 - \frac{1}{l})^2}}, \tag{2.26}$$

where Z_α is the α percentile for the standard Gaussian distribution.

2.7 Setting Counter Length

So far, our analysis has assumed that each counter has a sufficient number of bits such that it will not overflow. However, in order to save space, we want to set the counter length as short as possible. Suppose each measurement period ends after a pre-specified number n of packets are received. (Note that the value of n is the combined sizes of all flows during each measurement period.)

The average value of all counters will be $\frac{n}{m}$. We set the number of bits in each counter, denoted as b, to be $\log_2 \frac{n}{m} + 1$. Due to the additional bit, each counter can hold at least two times of the average before overflowing. If the allocated memory has M bits, the values of b and m can be determined from the following equations:

$$b \times m = M, \quad \log_2 \frac{n}{m} + 1 = b. \tag{2.27}$$

Due to the randomized counter sharing design, roughly speaking, the packets are distributed in the counters at random. In our experiments, the counter values approximately follow a Gaussian distribution with a mean of $\frac{n}{m}$. In this distribution, the fraction of counters that are more than four times of the mean is very small—less than 5.3 % in all the experiments. Consequently, the impact of counter overflow in CSM or MLM is also very small for most flows. Though it is small, we will totally eliminate this impact later in Sect. 2.10.4.

2.8 Flow Labels

The compact online data structure introduced in Sect. 2.4 only stores the flow size information. It does not store the flow labels. The labels are per-flow information, and it cannot be compressed in the same way we do for the flow sizes. In some applications, the flow labels are pre-known and do not have to be collected. For example, if an ISP wants to measure the traffic from its customers, it knows their IP addresses (which are the flow labels in this case). Similarly, if the system administrator of a large enterprise network needs the information about the traffic volumes of the hosts in the network, she has the hosts' addresses.

In case that the flow labels need to be collected and there is not enough SRAM to keep them, the labels have to be stored in DRAM. An efficient solution for label collection was proposed in [22]. A Bloom filter [2, 3] can be implemented in SRAM to encode the flow labels that have seen by the router during a measurement period, such that each label is only stored once in DRAM when it appears for the first time in the packet stream; storing each label once is the minimum overhead if the labels must be collected.

If we use three hash functions in the Bloom filter, each packet incurs three SRAM accesses in order to check whether the flow label carried the packet is already encoded in the Bloom filter. A recent work on one-memory-access Bloom filters [23] shows that three SRAM accesses per packet can be reduced to one. This overhead is further reduced if we only examine the UDF packets and the SYN packets (which carry the label information of TCP traffic). A recent study shows that UDF accounts for 20 % of the Internet traffic [4] and the measurement of our campus traffic shows that SYN packets accounts for less than 10 % of all TCP traffic. Therefore, the Bloom filter operation only needs to be carried out for less than 28 % of all packets, which amortizes the overhead.

2.9 Experiments

We use experiments to evaluate the estimation methods, CSM (Counter Sum estimation Method) and MLM (Maximum Likelihood estimation Method), which are designed based on the randomized counter sharing scheme. We also compare our

Table 2.1 Number of memory accesses and number of hash computations per packet

	Memory accesses	Hash operations	Constant?
CSM	2	1	Y
MLM	2	1	Y
CB	≥ 6	≥ 3	N
MRSCBF	4.47	4.47	N

methods with CB (Counter Braids) [21] and MRSCBF (Multi-Resolution Space-Code Bloom Filters) [19]. The evaluation is based on the performance metrics outlined in Sect. 2.1, including per-packet processing time, memory overhead, and estimation accuracy.

The experiments use a network traffic trace obtained from the main gateway of our campus. We perform experiments on various different types of flows, such as per-source flows, per-destination flows, per-source/destination flows, and TCP flows. They all lead to the same conclusions. Without losing generality, we choose TCP flows for presentation. The trace contains about 68 millions of TCP flows and 750 millions of packets. In each measurement period, 10 million packets are processed; it typically covers slightly more than 1 million flows.

2.9.1 Processing Time

The processing time is mainly determined by the number of memory accesses and the number of hash computations per packet. Table 2.1 presents the comparison. CSM or MLM performs two memory accesses and one hash computation for each packet. CB incurs three times of the overhead. It performs six memory accesses and three hash computations for each packet at the first counter level, and in the worst case makes six additional memory accesses and three additional hash computations at the second level. MRSCBF has nine filters. The ith filter uses k_i hash functions and encodes packets with a sampling probability p_i, where $k_1 = 3$, $k_2 = 4$, $k_i = 6$, $\forall i \in [3, 9]$, and $p_i = (\frac{1}{4})^{i-1}$, $\forall i \in [1, 9]$. When encoding a packet, the ith filter performs k_i hash computations and sets k_i bits. Hence, the total number of memory accesses (or hash computations) per packet for all filters is $\sum_{i=1}^{9}(p_i \cdot k_i) \approx 4.47$.

2.9.2 Memory Overhead and Estimation Accuracy

We provide the estimation accuracies of CSM and MLM under different levels of memory availability. In each measurement period, 10 M packets are processed, i.e., $n = 10$ M, which translates into about 8 s for an OC-192 link (10+ Gbps) or about 2 s for an OC-768 link (40+ Gbps) with an average packet size of 1,000 bytes.

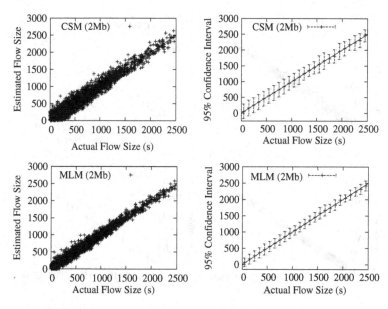

Fig. 2.2 Estimation results by CSM and MLM when $M = 2\,\mathrm{Mb}$

The memory M allocated to this particular measurement function is varied from $2\,\mathrm{Mb}$ (2×2^{20} bits) to $8\,\mathrm{Mb}$. The counter length b and the number of counters m are determined based on (2.27). The size of each storage vector is 50.

When $M = 2\,\mathrm{Mb}$, the experimental results are presented in Fig. 2.2. The top left plot shows the estimation results by CSM for one measurement period; the results for other measurement periods are very similar. Each flow is represented by a point in the plot, whose x coordinate is the true flow size s and y coordinate is the estimated flow size $\hat{s}$. The equality line, $y = x$, is also shown for reference. An estimation is more accurate if the point is closer to the equality line.

The top right plot presents the 95 % confidence intervals for the estimations made by CSM. The width of each vertical bar shows the size of the confidence interval at a certain flow size (which is the x coordinate of the bar). The middle point of each bar shows the mean estimation for all flows of that size. Intuitively, the estimation is more accurate if the confidence interval is smaller and the middle point is closer to the equality line.

The bottom left plot shows the estimation results by MLM, and the bottom right plot shows the 95 % confidence intervals for the estimations made by MLM. Clearly, MLM achieves better accuracy than CSM. The estimation accuracy shown in Fig. 2.2 is achieved with a memory of slightly less than 2 bits per flow.

We can improve the estimation accuracy of CSM or MLM by using more memory. We increase M to $4\,\mathrm{Mb}$ and repeat the above experiments. The results are shown in Fig. 2.3. We then increase M to $8\,\mathrm{Mb}$ and repeat the above experiments. The results

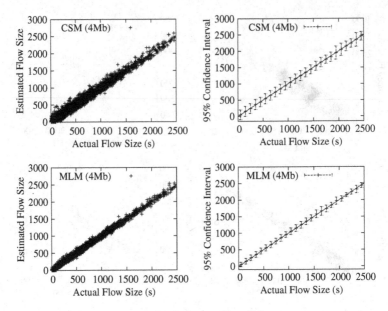

Fig. 2.3 Estimation results by CSM and MLM when $M = 4$ Mb

are shown in Fig. 2.4. The accuracy clearly improves as the confidence intervals shrink when M becomes larger.

We repeat the same experiments on CB, whose parameters are selected according to [21]. The results are presented in Fig. 2.5. The top left plot shows that CB totally fails to produce any meaningful results when the available memory is too small: $M = 2$ Mb, which translates into less than 2 bits per flow. In fact, its algorithm cannot converge, but instead produce oscillating results. We have to artificially stop the algorithm after a very long time. The top right plot shows that CB works well when $M = 4$ Mb. The algorithm still cannot converge by itself, even though it can produce very good results when we artificially stop it after a long time without observing any further improvement in the results. It can be seen that the results carry a small positive bias because most points are on one side of the equality line. The bottom plot shows that CB is able to return the exact sizes for most flows when the memory is $M = 8$ Mb.

Combining the results in Table 2.1, we draw the following conclusion: (1) In practice, we should choose CSM/MLM if the requirement is to handle high measurement throughput (which means low per-packet processing time) or if the available memory is too small such that CB does not work, while relatively coarse estimation is acceptable. (2) We should choose CB if the processing time is less of a concern, sufficient memory is available, and the exact flow sizes are required.

We also run MRSCBF under different levels of memory availability. We begin with $M = 8$ Mb. CSM or MLM works very well with this memory size (Fig. 2.4). The performance of MRSCBF is shown in the top left plot of Fig. 2.6. There are some very

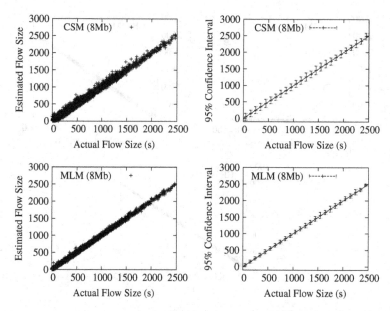

Fig. 2.4 Estimation results by CSM and MLM when $M = 8\,\mathrm{Mb}$

large estimated sizes. To control the scale in the vertical axis, we artificially set any estimation beyond 2,800 to be 2,800. The results demonstrate that MRSCBF totally fails when $M = 8\,\mathrm{Mb}$. The performance of MRSCBF improves when we increase the memory. The results when $M = 40\,\mathrm{Mb}$ are shown in the top right plot.[1] In the bottom left plot, when we further increase M to $80\,\mathrm{Mb}$,[2] no obvious improvement is observed when comparing with the top right plot. A final note is that the original work of MRSCBF uses log scale in their presentation. The bottom left plot in Fig. 2.6 will appear as the bottom right plot in log scale.

Clearly, the bitmap-based MRSCBF performs worse than CB, CSM or MLM. To measure flow sizes, counters are superior than bitmaps.

2.10 Extension of Estimation Range

We set the upper bound on the flow size that CSM and MLM can estimate in Sect. 2.9 to 2,500. However, in today's high-speed networks, the sizes of some flows are much larger than 2,500. In order to extend the estimation range to cover these large flows, we present four approaches that increase the estimation upper bound, and

---

[1] At the end of each measurement period, about half of the bits in the filters of MRSCBF are set to ones.

[2] At the end of each measurement period, less than half of the bits in the filters of MRSCBF are set to ones.

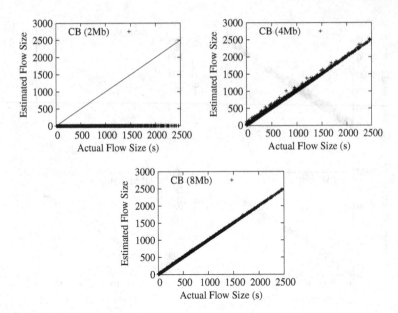

Fig. 2.5 Estimation results by CB when $M = 2, 4$, and $8\,M$

present extensive experimental results to demonstrate their effectiveness. Since MLM generally performs better than CSE, we only discuss how to extend the estimation range of MLM. CSE can be easily enhanced by similar approaches.

According to Sect. 2.4, each flow is assigned a unique storage vector. A flow's storage vector consists of l counters and each counter has b bits. Therefore, the maximum number of packets that the storage vector can represent is $l \times (2^b - 1)$. If we increase b by one, the number of packets that the vector can represent will be doubled. Similarly, if we increase l by a certain factor, the number of packets that the vector can represent will be increased by the same factor. Based on these observations, we extend the estimation range of MLM by increasing the value of b and l, respectively. In addition, we add a sampling module to MLM and consider hybrid SRAM/DRAM implementation to extend the estimation range.

2.10.1 Increasing Counter Size b

The first approach to extend the estimation range is to enlarge the counter size b. We repeat the same experiment on MLM presented in the bottom left plot of Fig. 2.3 (Sect. 2.9.2), where $M = 4\,\mathrm{Mb}$, $l = 50$, and $n = 10\,\mathrm{M}$. This time, instead of computing b from (2.27), we vary its value from 6 to 9. The new experimental results are shown in Fig. 2.7. In the top left plot, the maximum flow size that MLM can estimate is about 1,400 when $b = 6$. In the top right plot, where $b = 7$, the

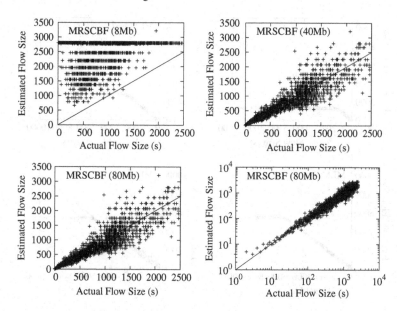

Fig. 2.6 Estimation results by MRSCBF

maximum flow size is about 2,800, which is twice of the maximum flow size that the top left plot can achieve. When b is set to 8, the bottom left plot shows that the estimation range of MLM is further extended. The bottom right plot shows that, when $b = 9$, the maximum flow size that MLM can estimate does not increase any more when comparing with the bottom left plot, which we will explain shortly.

The estimation accuracy of the above experiments is presented in Fig. 2.8, where the first plot shows the estimation bias and the second plot shows the standard deviation of the experimental results in Fig. 2.7. Generally speaking, both bias and standard deviation increase slightly when b increases.

Since flows share counters in MLM, the size information of one flow in a counter is the noise to other flows that share the same counter. When the amount of memory allocated to MLM is fixed ($M = 4$ Mb in these experiments), a larger value for b will result in a smaller value for m, i.e., the total number of counters is reduced. Hence, each counter has to be shared by more flows, and the average number of packets stored in each counter will increase. That means heavier noise among flows, which degrades the estimation accuracy, as is demonstrated by Fig. 2.8. Moreover, although a counter with a larger size b can keep track of a larger number of packets, since it also carries more noise, MLM has to subtract more noise from the counter value during the estimation process. As a result, the estimation range cannot be extended indefinitely by simply increasing b, which explains the fact that the maximum flow size that MLM can estimate does not increase when b reaches 9 in Fig. 2.7.

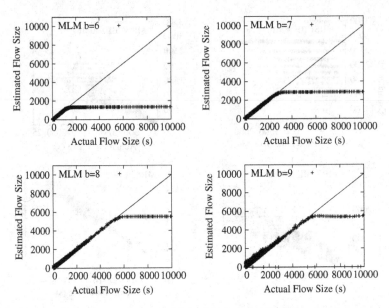

Fig. 2.7 Estimation results by MLM when $b = 6, 7, 8$, and 9. In these experiments, $n = 10\,M$, $M = 4\,Mb$

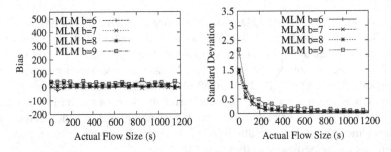

Fig. 2.8 Estimation bias and standard deviation in the experimental results shown in Fig. 2.7

2.10.2 Increasing Storage Vector Size l

The second approach for extending the estimation range is to increase the storage vector size l. We repeat the experiments in the previous subsection for MLM with $M = 4\,Mb$, $b = 7$, and $n = 10M$. We vary l from 50 to 1,000. Figure 2.9 presents the experimental results. The top left plot shows that the maximum flow size that MLM can estimate is about 5,800 when $l = 50$. As we increase the value of l, MLM can estimate increasingly larger flow sizes. However, when l becomes too large, estimation accuracy will degrade, which is evident in the bottom right plot. The reason is that each flow shares too many counters with others, which results

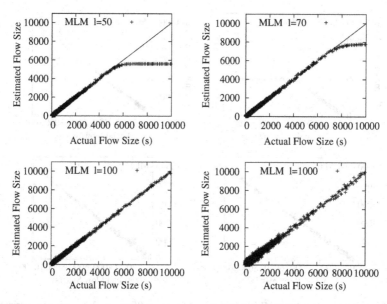

Fig. 2.9 Estimation results by MLM when $l = 50, 70, 100$, and $1,000$. In these experiments, $n = 10\,\text{M}$, $M = 4\,\text{M}$

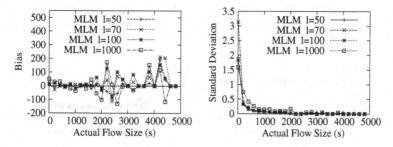

Fig. 2.10 Estimation bias and standard deviation in the experimental results shown in Fig. 2.9

in excessive noise in the counters and consequently introduce inaccuracy in the estimation process.

The estimation accuracy of the above experiments is presented in Fig. 2.10, where the first plot shows the estimation bias and the second plot shows the standard deviation of the experimental results in Fig. 2.9. Generally speaking, both bias and standard deviation increase slightly when l increases. Clearly, the value of l should not be chosen too large (such as $l = 1,000$) in order to prevent estimation accuracy to degrade significantly.

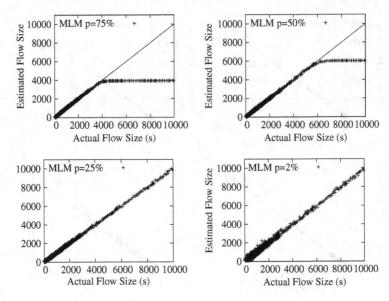

Fig. 2.11 Estimation results by MLM when $p = 75\%$, 50%, 25%, and 2%. In these experiments, $n = 10\,\text{M}$, $M = 4\,\text{Mb}$

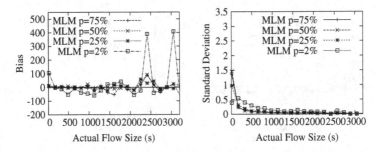

Fig. 2.12 Estimation bias and standard deviation in the experimental results shown in Fig. 2.11

2.10.3 Employing Sampling Module

In the third approach, we add a sampling module to MLM to enlarge the estimation range. The sampling technique has been widely used in network measurement [5, 10, 11, 19, 30]. We show that it also works for MLM. Let p be the sampling probability. For each packet that the router receives in the data encoding phase, the router generates a random number r in a range $[0, N]$. If $r < p \times N$, the router processes the packet as we describe in Sect. 2.4. Otherwise, it ignores the packet without encoding it in the counter array. In the data decoding phase, the estimated flow size should be $\frac{\hat{s}}{p}$, where $\hat{s}$ is computed from (2.18). The estimation range is expanded by a factor of $\frac{1}{p}$.

We again repeat the experiments in the previous sections for MLM with $M = 4\,\text{Mb}$, $l = 50$, and $n = 10\,\text{M}$. The value of b is computed from (2.27). This time, we introduce a sampling probability p and varies its value. Figure 2.11 presents the experimental results of MLM with $p = 75\,\%, 50\,\%, 25\,\%$, and $2\,\%$, respectively. It demonstrates that when the sampling probability decreases, the estimation range increases. However, it comes with a penalty on estimation accuracy. Figure 2.12 shows the estimation bias and standard deviation of the estimation results in Fig. 2.11. If the sampling probability is not decreased too small, e.g., when $p \geq 25\,\%$, the increase in bias and standard deviation is insignificant. However, if the sampling probability becomes too small such as $2\,\%$, the degradation in estimation accuracy also becomes noticeable.

2.10.4 Hybrid SRAM/DRAM Design

Can we extend the estimation range without any limitation and do so without any degradation in estimation accuracy? This will require a hybrid SRAM/DRAM design. In SRAM, we still choose the value of b based on (2.27). The limited size of each counter means that a counter may be overflowed during the data encoding phase even though the chance for this to happen is very small (Sect. 2.7). To totally eliminate the impact of counter overflow, we keep another array of counters in DRAM, each of which has a sufficient number of bits. The counters in DRAM are one-to-one mapped to the counters in SRAM. When a counter in SRAM is overflowed, it is reset to zero and the corresponding counter in DRAM is incremented by one. During offline data analysis, the counter values are set based on both SRAM and DRAM data. Because overflow happens only to a small fraction of SRAM counters and a DRAM access is made only after an overflowed SRAM counter is accessed 2^b times, the overall overhead of DRAM access is very small.

2.11 Summary

Per-flow traffic measurement provides real-world data for a variety of applications on accounting and billing, anomaly detection, and traffic engineering. Online data collection methods must meet the requirements of being both fast and compact. This chapter presents a novel data encoding/decoding scheme, which mixes per-flow information randomly in a tight SRAM space for compactness. Its online operation only incurs a small overhead of one hash computation and one counter update per packet. Two offline statistical methods—the counter sum estimation and the maximum likelihood estimation—are used to extract per-flow sizes from the mixed data structures with good accuracy. Due to its design philosophy that is fundamentally different from the prior art, the new measurement function is able to work in a tight space where exact measurement is no longer possible, and it does so with the minimal number of memory accesses per packet.

References

1. Abramowitz, M., Stegun, I.: Handbook of Mathematical Functions: with Formulas, Graphs, and Mathematical Tables. Dover Publications, New York (1964)
2. Bloom, B.H.: Space/Time trade-offs in hash coding with allowable errors. Commun. ACM **13**(7), 422–426 (1970)
3. Broder, A., Mitzenmacher, M.: Network applications of bloom filters: a survey. Internet Math. **1**(4), 485–509 (2002)
4. CAIDA: Analyzing UDP Usage in Internet Traffic. http://www.caida.org/research/traffic-analysis/tcpudpratio/ (2009)
5. Cao, J., Jin, Y., Chen, A., Bu, T., Zhang, Z.: Identifying high cardinality internet hosts. In: Proceedings of IEEE INFOCOM (2009)
6. Casella, G., Berger, R.: Statistical Inference. Duxbury Press, Pacific Grove (2001)
7. Cohen, S., Matias, Y.: Spectral bloom filters. In: Proceedings of ACM SIGMOD (2003)
8. Cvetkovski, A.: An algorithm for approximate counting using limited memory resources. Proceedings of ACM SIGMETRICS (2007)
9. Demaine, E., Lopez-Ortiz, A., Munro, J.: Frequency estimation of internet packet streams with limited space. In: Proceedings of 10th ESA Annual European Symposium on Algorithms (2002)
10. Dimitropoulos, X., Hurley, P., Kind, A.: Probabilistic lossy counting: an efficient algorithm for finding heavy hitters. ACM SIGCOMM Comput. Comm. Rev. **38**(1), 7–16 (2008)
11. Duffield, N., Lund, C., Thorup, M.: Estimating flow distributions from sampled flow statistics. In: Proceedings of ACM SIGCOMM (2003)
12. Estan, C., Varghese, G.: New directions in traffic measurement and accounting. In: Proceedings of ACM SIGCOMM (2002)
13. Estan, C., Varghese, G., Fish, M.: Bitmap algorithms for counting active flows on high-speed links. IEEE/ACM Trans. Netw. (TON) **14**(5), 925–937 (2006)
14. Hao, F., Kodialam, M., Lakshman, T.V.: ACCEL-RATE: A faster mechanism for memory efficient per-flow traffic estimation. In: Proceedings of ACM SIGMETRICS/Performance (2004)
15. Kamiyama, N., Mori, T.: Simple and accurate identification of high-rate flows by packet sampling. In: Proceedings of IEEE INFOCOM (2006)
16. Karp, R., Shenker, S., Papadimitriou, C.: A simple algorithm for finding frequent elements in streams and bags. ACM Trans. Database Syst. **28**(1), 51–55 (2003)
17. Kodialam, M., Lakshman, T.V., Mohanty, S.: Runs based traffic estimator (RATE): a simple, memory efficient scheme for per-flow rate estimation. In: Proceedings of IEEE INFOCOM (2004)
18. Kumar, A., Sung, M., Xu, J., Wang, J.: Data streaming algorithms for efficient and accurate estimation of flow size distribution. In: Proceedings of ACM SIGMETRICS (2004)
19. Kumar, A., Xu, J., Wang, J., Spatschek, O., Li, L.: Space-code bloom filter for efficient per-flow traffic measurement. In: Proceedings of IEEE INFOCOM (2004)
20. Lehmann, E., Casella, G.: Theory of Point Estimation. Springer, New York (1998)
21. Lu, Y., Montanari, A., Prabhakar, B., Dharmapurikar, S., Kabbani, A.: Counter braids: a novel counter architecture for per-flow measurement. In: Proceedings of ACM SIGMETRICS (2008)
22. Lu, Y., Prabhakar, B.: Robust Counting via counter braids: an error-resilient network measurement architecture. In: Proceedings of IEEE INFOCOM (2009)
23. Qiao, Y., Chen, S., Li, T., Chen, S.: Energy-efficient polling protocols in RFID systems. In: Proceedings of ACM MobiHoc (2011)

24. Ramakrishna, M., Fu, E., Bahcekapili, E.: Efficient hardware hashing functions for high performance computers. IEEE Trans. Comput. **46**(12), 1378–1381 (1997)
25. Stanojevic, R.: Small active counters. In: Proceedings of IEEE INFOCOM (2007)
26. Venkatataman, S., Song, D., Gibbons, P., Blum, A.: New streaming algorithms for fast detection of superspreaders. In: Proceedings of NDSS (2005)
27. Yoon, M., Li, T., Chen, S., Peir, J.: Fit a spread estimator in small memory. In: Proceedings of IEEE INFOCOM (2009)
28. Yoon, M., Li, T., Chen, S., Peir, J.: Fit a compact spread estimator in small high-speed memory. IEEE/ACM Trans. Netw. **19**(5), 1253–1264 (2011)
29. Zhao, H., Wang, H., Lin, B., Xu, J.: Design and performance analysis of a DRAM-based statistics counter array architecture. In: Proceedings of ACM/IEEE ANCS (2009)
30. Zhao, Q., Kumar, A., Xu, J.: Joint data streaming and sampling techniques for detection of super sources and destinations. In: Proceedings of USENIX/ACM Internet Measurement Conference (2005)

Chapter 3
Spreader Classification

Abstract This chapter discusses the problem of spreader classification. It provides an efficient spread estimator based on dynamic bit sharing, which incorporates probabilistic sampling and bit sharing for compact information storage. The estimator applies a maximum likelihood estimation method to extract per-source information from the shared bits in order to determine the heavy spreaders. It ensures that the false positive/false negative ratios are bounded with high probability. Moreover, given an arbitrary set of bounds, the chapter develops a systematic approach to determine the optimal system parameters that minimize the amount of memory needed to meet the bounds. Experiments based on a real Internet traffic trace demonstrate that this spreader classification scheme reduces memory consumption by three to twenty times when comparing with related work. The rest of this chapter is organized as follows: Sect. 3.1 gives the problem definition. Section 3.2 presents an efficient spreader classification scheme. Section 3.3 presents the analytical results for optimal parameters. Section 3.4 presents the experimental results. Section 3.5 describes a multi-objective spreader classification problem and the solution. Section 3.6 describes other methods. Section 3.7 gives the summary.

Keywords Spread estimator · Dynamic bit sharing

3.1 Problem Statement

How to formally define the classification objective? A straightforward objective is to report all sources whose spread values exceed a certain threshold. However, unless we measure the spread of each source precisely, we cannot accurately classify sources based on the threshold. Precise spread measurement is a costly operation, and most existing work resorts to spread estimation. Naturally, the follow-up question is how to define a classification objective that embodies a probabilistic performance bound.

T. Li and S. Chen, *Traffic Measurement on the Internet,*
SpringerBriefs in Computer Science, DOI: 10.1007/978-1-4614-4851-8_3,
© The Author(s) 2012

We adopt the *probabilistic classification objective* from [14]. Let h and l be two positive integers, $h > l$. Let α and β be two probability values, $0 < \alpha < 1$ and $0 < \beta < 1$. The objective is to report any source whose spread is h or larger with a probability no less than α and report any source whose spread is l or smaller with a probability no more than β. Let k be the spread of an arbitrary source src. The objective can be expressed in terms of conditional probabilities:

$$Prob\{\text{report } src \text{ as a heavy spreader} \mid k \geq h\} \geq \alpha \qquad (3.1)$$
$$Prob\{\text{report } src \text{ as a heavy spreader} \mid k \leq l\} \leq \beta$$

We treat the report of a source whose spread is l or smaller as a *false positive*, and the non-report of a source whose spread is h or larger as a *false negative*. Hence, the above objective can also be stated as bounding the false positive ratio by β and the false negative ratio by $1 - \alpha$. The goal is to minimize the amount of SRAM that is needed for achieving the above objective.

Although the technical discussion in this chapter focuses on spreader classification of sources, the same techniques can be equally applied to spreader classification of destinations.

3.2 An Efficient Spreader Classification Scheme

This section presents an efficient spreader classification scheme (ESC), which is the combination of probabilistic sampling, dynamic bit sharing, and maximum likelihood estimation.

3.2.1 Probabilistic Sampling

To save space, a router samples the contacts made by external sources to internal destinations, and it only stores the sampled contacts. The router selects contacts for storage uniformly at random with a sampling probability p. The sampling procedure is simple: the router hashes the source/destination address pair of each packet that arrives at the external network interface into a number in a range $[0, N)$. If the hash result is smaller than $p \times N$, the contact will be stored; otherwise, the contact will not be stored.

3.2.2 Bit-Sharing Storage

For each source, a bit vector (also called *bitmap*) may be used to store all its sampled contacts. The bits are initially zeros. Each sampled contact is hashed to a bit in the bitmap, and the bit is set to one. At the end of the measurement period, using the method of probabilistic counting [12] or its variants [4, 10], we can estimate the number of contacts (i.e., the spread of the source) based on the number of zeros remaining in the bitmap.

However, using per-source bitmaps is not memory-efficient. If each bitmap is 32 bits long and there are 1M sources with sampled contacts, the total memory requirement will be 32 Mb. If the allocated SRAM space is much smaller, e.g., 0.5 Mb, there will not be enough bitmaps for all sources. This problem cannot be solved simply by reducing the size of each bitmap because even if a bitmap has just one bit, it still takes 1 Mb. Moreover, the performance of probabilistic counting [12] or its variants [4, 10] requires bitmaps to be not too small.

Our solution is to mix all bitmaps together and let them share bits, such that an almost arbitrary number of bitmaps can be created from a limited available space. Bit sharing among bitmaps causes information interference, which will be removed when we derive our formula. The level of bit sharing, which is determined by system parameters (see the next section), controls the tradeoff between classification accuracy and space overhead. Details of this method is presented below.

Let m be the total number of available bits. All bits are organized in a single array B. For an arbitrary source src, we use a hash function to pseudo-randomly select a number of bits from B to store the contacts made by src. The indices of the selected bits are $H(src \oplus R[0])$, $H(src \oplus R[1])$, ..., $H(src \oplus R[s-1])$, where $H(...)$ is a hash function whose range is $[0, m)$, R is an integer array, storing randomly chosen constants whose purpose is to arbitrarily alter the hash result, and s ($\ll m$) is a system parameter that specifies the number of bits to be selected. The above bits form a *logical bitmap* of source src, denoted as $LB(src)$.

Similarly, a logical bitmap can be constructed from B for any other source. Essentially, we embed the bitmaps of all possible sources in B. The bit-sharing relationship is dynamically determined on the fly as new sources are allocated logical bitmaps from B.

At the beginning of a measurement period, all bits in B are reset to zeros. Consider an arbitrary contact $\langle src, dst \rangle$ that is sampled for storage, where src is the source address and dst is the destination address. The router sets *a single bit* in B to one. Obviously, it must also be a bit in the logical bitmap $LB(src)$. The index of the bit to be set for this contact is given as follows:

$$H(src \oplus R[H(dst \oplus K) \bmod s]).$$

The outer hash, $H(src \oplus R[...])$, ensures that it is a bit in $LB(src)$. The inner hash, $H(dst \oplus K)$, ensures that the bit is pseudo-randomly selected from $LB(src)$. The private key K is introduced to prevent the *hash collision attacks*. In such an attack, a heavy spreader src finds a set of destination addresses, $dst_1, dst_2, ...,$ that have the same hash value, $H(dst_1) = H(dst_2) = ...$ If it only contacts these destinations, the same bit in $LB(src)$ will be set, which allows the heavy spreader to stay undetected. This type of attacks can be prevented if we use a cryptographic hash function such as MD5 or SHA1, which makes it difficult to find destination addresses that have the same hash value. However, if a weaker hash function is used for performance reason, then a private key becomes necessary. Without knowing the key, the heavy spreaders will not be able to predict which destination addresses produce the same hash value.

To store a contact, ESC only sets a single bit. This is more efficient than other methods [14, 18] that require setting multiple bits for storing each contact.

3.2.3 *Maximum Likelihood Estimation and Heavy Spreader Classification*

At the end of the measurement period, ESC will send the content of B to an offline data processing center. There, the logical bitmap of each source src is extracted and the estimated spread $\hat{k}$ of the source is computed. Only if $\hat{k}$ is greater than a threshold value T, ESC reports the source as a heavy spreader. We will discuss how to keep track of the source addresses and explain how to determine the threshold T based on a given classification objective later. The formula for computing the estimated spread $\hat{k}$ is

$$\hat{k} = \frac{\ln V_s - \ln V_m}{\ln(1 - \frac{p}{s}) - \ln(1 - \frac{p}{m})}, \tag{3.2}$$

where p is the sampling probability, s is the size of the logical bitmap $LB(src)$, m is the size of B, V_s is the fraction of bits in $LB(src)$ whose values are zeros, and V_m is the fraction of bits in B whose values are zeros. Below we formally derive this formula.

Let k be the true spread of source src, n be the number of distinct contacts made by all sources, and U_s be the number of bits in $LB(src)$ whose values are zeros. Clearly, $V_s = \frac{U_s}{s}$. Depending on the context, U_s (or V_s, V_m) is used either as a random variable or an instance value of the random variable.

The probability for any contact to be sampled for storage is p. Consider an arbitrary bit b in $LB(src)$. A sampled contact made by src has a probability of $\frac{1}{s}$ to set b to '1', and a sampled contact made by any other source has a probability of $\frac{1}{m}$ to set b to '1'. Hence, the probability $q(k)$ for b to remain '0' at the end of the measurement period is

$$q(k) = (1 - \frac{p}{m})^{n-k}(1 - \frac{p}{s})^k. \tag{3.3}$$

Each bit in $LB(src)$ has a probability of $q(k)$ to remain '0'. The observed number of '0' bits in $LB(src)$ is U_s. The likelihood function for this observation to occur is given as follows:

$$L = q(k)^{U_s}(1 - q(k))^{s-U_s}. \tag{3.4}$$

In the standard process of maximum likelihood estimation, the unknown value k is technically treated as a variable in (3.4). We want to find an estimate $\hat{k}$ that maximizes the likelihood function. Namely,

$$\hat{k} = \arg\max_k\{L\}. \tag{3.5}$$

Since the maxima is not affected by monotone transformations, we use logarithm to turn the right side of (3.4) from product to summation:

$$\ln(L) = U_s \cdot \ln(q(k)) + (s - U_s) \cdot \ln(1 - q(k)).$$

From (3.3), the above equation can be written as

$$\ln(L) = U_s((n-k)\ln(1 - \frac{p}{m}) + k\ln(1 - \frac{p}{s}))$$
$$+ (s - U_s) \cdot \ln(1 - (1 - \frac{p}{m})^{n-k}(1 - \frac{p}{s})^k).$$

To find the maxima, we differentiate both sides:

$$\frac{\partial \ln(L)}{\partial k} = \ln(\frac{1 - \frac{p}{s}}{1 - \frac{p}{m}}) \cdot \frac{U_s - s(1 - \frac{p}{m})^{n-k}(1 - \frac{p}{s})^k}{1 - (1 - \frac{p}{m})^{n-k}(1 - \frac{p}{s})^k}. \tag{3.6}$$

We then let the right side be zero. That is,

$$U_s = s(1 - \frac{p}{m})^{n-k}(1 - \frac{p}{s})^k. \tag{3.7}$$

Taking logarithm on both sides, we have

$$\ln\frac{U_s}{s} = n\ln(1 - \frac{p}{m}) + k(\ln(1 - \frac{p}{s}) - \ln(1 - \frac{p}{m})),$$
$$k = \frac{\ln V_s - n\ln(1 - \frac{p}{m})}{\ln(1 - \frac{p}{s}) - \ln(1 - \frac{p}{m})}. \tag{3.8}$$

Suppose the number of sources (which equals to the number of logical bitmaps) is sufficiently large. Because every bit in every logical bitmap is randomly selected from B, in this sense, each of the n contacts has about the same probability $\frac{p}{m}$ of setting any bit in B. Hence, we have

$$E(V_m) = (1 - \frac{p}{m})^n. \tag{3.9}$$

Applying (3.9) to (3.8), we have

$$k = \frac{\ln V_s - \ln E(V_m)}{\ln(1 - \frac{p}{s}) - \ln(1 - \frac{p}{m})}. \tag{3.10}$$

Replacing $E(V_m)$ by the instance value V_m, we have the following estimation for k.

$$\hat{k} = \frac{\ln V_s - \ln V_m}{\ln(1 - \frac{p}{s}) - \ln(1 - \frac{p}{m})}, \tag{3.11}$$

where V_s can be measured by counting the number of zeros in $LB(src)$, V_m can be measured by counting the number of zeros in B, and s, p and m are pre-set parameters of ESC (see the next section).

We give an intuitive explanation for (3.11). First, we point out that its development has taken information interference (due to bit sharing) into consideration. In (3.3), the term $(1 - \frac{p}{m})^{n-k}$ captures the effect of interference; it is the probability that a bit in $LB(src)$ is not set by any contact from a different source. Eventually, this results in two terms, $\ln V_m$ and $\ln(1 - \frac{p}{m})$, in (3.11). Second, the formula (3.11) is an estimation. It does not give the precise value of k due to its probabilistic bit setting nature. It is even mathematically possible to give a negative estimation though this happens with exceedingly low probability. Our analysis will demonstrate that, as long as the system parameters are set such that (3.11) gives good estimations with high probability, the objective in (3.1) will be met.

3.2.4 Variance of V_m

Let A_i be the event that the ith bit in B remains '0' at the end of the measurement period and 1_{A_i} be the corresponding indicator random variable. Let U_m be the random variable for the number of '0' bits in B. We first derive the probability for A_i to occur and the expected value of U_m. For an arbitrary bit in B, each distinct contact has a probability of $\frac{p}{m}$ to set the bit to one. All contacts are independent of each other when setting bits in B. Hence,

$$Prob\{A_i\} = \left(1 - \frac{p}{m}\right)^n, \quad \forall i \in [0, s).$$

The probability for A_i and A_j, $\forall i, j \in [0, m), i \neq j$, to happen simultaneously is

$$Prob\{A_i \cap A_j\} = (1 - \frac{2p}{m})^n.$$

Since $V_m = \frac{U_m}{m}$ and $U_m = \sum_{i=1}^{m} 1_{A_i}$, we have

$$E(V_m^2) = \frac{1}{m^2} E\left(\left(\sum_{i=1}^{m} 1_{A_i}\right)^2\right)$$

$$= \frac{1}{m^2} E\left(\sum_{i=1}^{m} 1_{A_i}^2\right) + \frac{2}{m^2} E\left(\sum_{1 \leq i < j \leq m} 1_{A_i} 1_{A_j}\right)$$

$$= \frac{1}{m}(1 - \frac{p}{m})^n + \frac{m-1}{m}(1 - \frac{2p}{m})^n.$$

Based on (3.9) and the equation above, we have

$$Var(V_m) = E(V_m^2) - E(V_m)^2$$

$$= \frac{1}{m}\left(1 - \frac{p}{m}\right)^n + \frac{m-1}{m}\left(1 - \frac{2p}{m}\right)^n - \left(1 - \frac{p}{m}\right)^{2n}$$

$$\simeq \frac{e^{-\frac{np}{m}}\left(1 - \left(1 + \frac{np^2}{m}\right)e^{-\frac{np}{m}}\right)}{m}. \tag{3.12}$$

3.2.5 Source Addresses

ESC does not store the source address of every arrival packet. Instead, it stores a source address only when a contact sets a bit in B from '0' to '1'. The frequency of storing source addresses is by far smaller than the packet arrival rate due to the following reasons. First, numerous packets may be sent from a source to a destination in a TCP/UDP session. Only the first sampled packet may cause the source address to be stored because only the first packet sets a bit from '0' to '1' and the remaining packets will set the same bit (which is already '1'). Therefore, for any TCP/UDP session, no matter how many packets it has, it triggers source address storage at most once. Second, a source may send thousands or even millions of packets through a router, but the number of times its address will be stored is bounded by s (which is the number of bits in the source's logical bitmap). Third, each arrival packet has a probability of p to be sampled. When it is not sampled, it has no chance to trigger source address storage. In summary, because the operation of storing source addresses is relatively infrequent, these addresses can be stored in the main memory.

3.3 Optimal System Parameters and Minimum Memory Requirement

In this section, we first derive the probability for ESC to report an arbitrary source as a heavy spreader. Based on this probability, we develop the constraints that the system parameters must satisfy in order to achieve the classification objective in (3.1). Using these constraints, we determine the optimal values for the size s of the logical bitmaps, the sampling probability p, and the threshold T. We also determine the minimum amount of memory m that should be allocated for ESC to achieve the objective.

3.3.1 Report Probability

Consider an arbitrary source src. ESC reports src as a heavy spreader if its estimated spread $\hat{k}$ exceeds a threshold T. The probability for this to happen, $Prob\{\hat{k} \geq T\}$ is derived as follows: from (3.11), we know that the following inequalities are equivalent.

$$\hat{k} \geq T$$

$$\frac{\ln V_s - \ln V_m}{\ln(1 - \frac{p}{s}) - \ln(1 - \frac{p}{m})} \geq T$$

$$V_s \leq V_m \left(\frac{1 - \frac{p}{s}}{1 - \frac{p}{m}}\right)^T$$

U_s is the random variable for the number of '0' bits in $LB(src)$. $U_s = s \cdot V_s$. The above inequality becomes

$$U_s \leq s \cdot V_m \cdot \left(\frac{1 - \frac{p}{s}}{1 - \frac{p}{m}}\right)^T . \tag{3.13}$$

For a set of parameters m, s, p and T, we define a constant

$$C = s \cdot V_m \cdot \left(\frac{1 - \frac{p}{s}}{1 - \frac{p}{m}}\right)^T ,$$

where the instance value of V_m can be measured from B after the measurement period. Hence, the probability for ESC to report src is $Prob\{\hat{k} \geq T\} = Prob\{U_s \leq C\}$.

U_s follows the binomial distribution with parameters s and $q(k)$, where $q(k)$ in (3.3) is the probability for an arbitrary bit in $LB(src)$ to remain zero at the end of the measurement period. Hence, the probability of having exactly i zeros in $LB(src)$ is given by the following probability mass function:

$$Prob\{U_s = i\} = \binom{s}{i} \cdot q(k)^i \cdot (1 - q(k))^{s-i}. \tag{3.14}$$

We must have

$$Prob\{\hat{k} \geq T\} = Prob\{U_s \leq C\}$$

$$= \sum_{i=0}^{\lfloor C \rfloor} \binom{s}{i} \cdot q(k)^i \cdot (1 - q(k))^{s-i}. \tag{3.15}$$

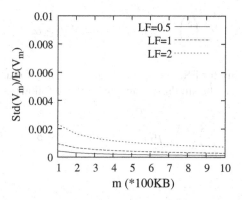

Fig. 3.1 The relative standard deviation, $\frac{Std(V_m)}{E(V_m)}$, approaches to zero as m increases. The *load factor* (LF) is defined as $n \cdot p/m$, where $n \cdot p$ is the number of distinct contacts that are sampled by ESC for storage. In our experiments (reported in Sect. 3.4), when we use the system parameters determined by the algorithm described in this section, the load factor never exceeds 2

3.3.2 Constraints for the System Parameters

We now derive the constraints that the system parameters must satisfy in order to achieve the classification objective in (3.1). The variance of V_m is given in (3.12). It approaches to zero as m increases. In Fig. 3.1, we plot the ratio of the standard deviation $Std(V_m) = \sqrt{Var(V_m)}$ to $E(V_m)$, which can be found in (3.9). The figure shows that $Std(V_m)/E(V_m)$ is very small when m is reasonably large. In this case, we can approximately treat V_m as a constant.

$$V_m \simeq E(V_m) \simeq \left(1 - \frac{p}{m}\right)^n. \tag{3.16}$$

The classification objective can be stated as two requirements. First, the probability for ESC to report a source with $k \geq h$ must be at least α. That is, $Prob\{\hat{k} \geq T\} \geq \alpha, \forall k \geq h$. From (3.15), this requirement can be written as the following inequality:

$$\sum_{i=0}^{\lfloor C \rfloor} \binom{s}{i} \cdot q(k)^i \cdot (1 - q(k))^{s-i} \geq \alpha,$$

where $C = s \cdot V_m \cdot (\frac{1-\frac{p}{s}}{1-\frac{p}{m}})^T \simeq s \cdot (1 - \frac{p}{m})^n \cdot (\frac{1-\frac{p}{s}}{1-\frac{p}{m}})^T$. The left side of the inequality is an increasing function in k. Hence, to satisfy the requirement in the worst case when $k = h$, the following constraint for the system parameters must be met:

$$\sum_{i=0}^{\lfloor C \rfloor} \binom{s}{i} \cdot q(h)^i \cdot (1 - q(h))^{s-i} \geq \alpha. \tag{3.17}$$

Second, the probability for ESC to report a source with $k \leq l$ must be no more than β. This requirement can be similarly converted into the following constraint:

$$\sum_{i=0}^{\lfloor C \rfloor} \binom{s}{i} \cdot q(l)^i \cdot (1 - q(l))^{s-i} \leq \beta. \tag{3.18}$$

3.3.3 Optimal System Parameters

Our goal is to optimize the system parameters such that the memory requirement, m, is minimized under the constraints (3.17) and (3.18). The problem is formally defined as follows.

$$Minimize \quad m \tag{3.19}$$

$$Subject \ to \quad \sum_{i=0}^{\lfloor C \rfloor} \binom{s}{i} \cdot q(h)^i \cdot (1 - q(h))^{s-i} \geq \alpha,$$

$$\sum_{i=0}^{\lfloor C \rfloor} \binom{s}{i} \cdot q(l)^i \cdot (1 - q(l))^{s-i} \leq \beta,$$

$$C = s \cdot \left(1 - \frac{p}{m}\right)^n \cdot \left(\frac{1 - \frac{p}{s}}{1 - \frac{p}{m}}\right)^T.$$

The parameters, h, l, α and β, are specified in the classification objective. The value of n is decided based on the history data in the past measurement periods. To be conservative, we take the maximum number n^* of distinct contacts observed in a number of previous measurement periods. More specifically (3.9) can be turned into a formula for estimating n in each previous period if we replace $E(V_m)$ with the instance value V_m.

$$\hat{n} = -\frac{m}{p} \ln V_m \tag{3.20}$$

We derive the relative bias and the relative standard deviation of the above estimation.

$$Bias\left(\frac{\hat{n}}{n}\right) = E\left(\frac{\hat{n}}{n}\right) - 1 \simeq \frac{e^{\frac{np}{m}} - \frac{np^2}{m} - 1}{2np} \tag{3.21}$$

$$Std\left(\frac{\hat{n}}{n}\right) = \frac{\sqrt{m}}{np}\left(e^{\frac{np}{m}} - \frac{np^2}{m} - 1\right)^{1/2} \tag{3.22}$$

They both approach to zero as m increases. Based on the largest $\hat{n}$ value observed in a certain number of past measurement periods, we can set the value of n^*.

To solve the constrained optimization problem (3.19), we need to determine the optimal values of the remaining three system parameters, s, p and T, such that m will be minimized. We consider the left side of (3.17) as a function $F_h(m, s, p, T)$, and the left side of (3.18) as $F_l(m, s, p, T)$. Namely,

$$F_h(m, s, p, T) = \sum_{i=0}^{\lfloor C \rfloor} \binom{s}{i} \cdot q(h)^i \cdot (1 - q(h))^{s-i},$$

$$F_l(m, s, p, T) = \sum_{i=0}^{\lfloor C \rfloor} \binom{s}{i} \cdot q(l)^i \cdot (1 - q(l))^{s-i}.$$

Both of them are *non-increasing functions in* T, according to the relation between C and T. In the following, we present an iterative numerical algorithm to solve the optimization problem. The algorithm consists of four procedures.

- First, we construct a procedure called $Potential(m, s, p)$, which takes a value of m, a value of s and a value of p as input and returns the maximum value of $F_h(m, s, p, T)$ under the condition that $F_l(m, s, p, T) \leq \beta$ is satisfied. Because $F_h(m, s, p, T)$ is a non-increasing function in T, we need to find the smallest value of T that satisfies $F_l(m, s, p, T) \leq \beta$. That can be done numerically through binary search: pick a small integer T_1 such that $F_l(m, s, p, T_1) \geq \beta$ and a large integer T_2 such that $F_l(m, s, p, T_2) \leq \beta$. We iteratively shrink the difference between them by resetting one of them to be the average $\frac{T_1+T_2}{2}$, while maintaining the inequalities, $F_l(m, s, p, T_1) \geq \beta$ and $F_l(m, s, p, T_2) \leq \beta$. The process stops when $T_1 = T_2$, which is denoted as T^*. The procedure $Potential(m, s, p)$ returns $F_h(m, s, p, T^*)$. The pseudo code is presented in Algorithm 1 in the appendix. Essentially, what $Potential(m, s, p)$ returns is the maximum value of the left side in (3.17) under the condition that (3.18) is satisfied. The difference between $Potential(m, s, p)$ and α provides us with a quantitative indication on how conservative or aggressive we have chosen the value of m. If $Potential(m, s, p) - \alpha$ is positive, it means that the performance achieved by the current memory size is more than required. We shall reduce m. On the contrary, if $Potential(m, s, p) - \alpha$ is negative, we shall increase m. Given the above semantics, when we determine the optimal values for p and s, our goal is certainly to maximize the return value of $Potential(m, s, p)$.
- Second, given a value of m and a value of s, we construct a procedure $Optimal$ $P(m, s)$ that determines the optimal value p^* such that $Potential(m, s, p^*)$ is maximized. When the values of m and s are fixed, $Potential(m, s, p)$ becomes a

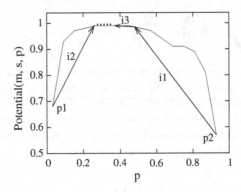

Fig. 3.2 a The *curve* (*without the arrows*) shows the value of $Potential(m, s, p)$ with respect to p when $m = 0.45$ MB and $s = 150$. Its non-smooth appearance is due to $\lfloor C \rfloor$ in the formula of $F_h(m, s, p, T^*)$. $F_h(m, s, p, T^*)$ depends on the values of $\lfloor C \rfloor$ and $q(h)$, which are both functions of p. **b** The *arrows* illustrate the operation of $Optimal P(m, s)$. In the first iteration (*arrow i_1*), p_2 is set to be $(p_1 + p_2)/2$. In the second iteration (*arrow i_2*), p_1 is set to be $(p_1 + p_2)/2$. In the third iteration (*arrow i_3*), p_2 is set to be $(p_1 + p_2)/2$

function of p. It is a curve as illustrated in Fig. 3.2; see explanation under caption (A) and ignore the arrows in the figure for now.

We use a binary search algorithm to find a near-optimal value of p. Let $p_1 = 0$ and $p_2 = 1$. Let δ be a small positive value (such as 0.001). Repeat the following operation: let $\bar{p} = (p_1 + p_2)/2$. If $Potential(m, s, \bar{p}) < Potential(m, s, \bar{p} + \delta)$, set p_1 to be $\bar{p}$; otherwise, set p_2 to be $\bar{p}$. The above iterative operation stops when $p_2 - p_1 < \delta$. The procedure $Optimal P(m, s)$ returns $(p_1 + p_2)/2$, which is within $\pm \delta/2$ of the optimal. This difference can be made arbitrarily small when we decrease δ at the expense of increased computation overhead. We want to stress that it is one-time overhead (not online overhead) to determine the system parameters before deployment. The operation of $Optimal P(m, s)$ is illustrated by the arrows in Fig. 3.2; see explanation under caption (B). The pseudo code is given in Algorithm 2 in the appendix.

- Third, given a value of m, we construct a procedure $Optimal S(m)$ that determines the optimal value s^* such that $Potential(m, s^*, Optimal P(m, s^*))$ is maximized. When the value of m is fixed, $Potential(m, s, Optimal P(m, s))$ becomes a function of s. It is a curve as illustrated in Fig. 3.3. We can use a binary search algorithm similar to that of $Optimal P(m, s)$ to find s^*. The pseudo code is given in Algorithm 3 in the appendix.

- Fourth, we construct a procedure $Optimal M()$ that determines the minimum memory requirement m^* through binary search: denote $Potential(m, Optimal S(m), Optimal P(m, Optimal S(m)))$ as $Potential(m, ...)$. Pick a small value m_1 such that $Potential(m_1, ...) \leq \alpha$, which means that the classification objective is not met — more specifically, according to the semantics of $Potential(...)$, the constraint (3.17) cannot be satisfied if the constraint (3.18) is satisfied. Pick a large value m_2 such that $Potential(m_2, ...) \geq \alpha$, which means that the classification

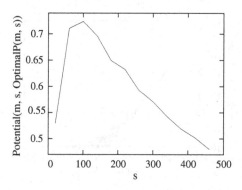

Fig. 3.3 The value of $Potential(m, s, OptimalP(m, s))$ with respect to s when $m = 0.25\,\text{MB}$

objective is met. Repeat the following operation. Let $\bar{m} = \lfloor (m_1 + m_2)/2 \rfloor$. If $Potential(\bar{m}, ...) \leq \alpha$, set m_1 to be $\bar{m}$; otherwise, set m_2 to be $\bar{m}$. The above iterative operation terminates when $m_1 = m_2$, which is returned by the procedure $OptimalM()$. The pseudo code is given in Algorithm 4 in the appendix.

In practice, a network administrator will first define the classification objective that is specified by α, β, h and l. He or she sets the value of n^* based on historic data, and then sets $m = OptimalM()$, $s = OptimalS(m)$, $p = OptimalP(m, s)$ and T as the threshold value T^* before the last call to $Potential(m, s, p)$ is returned during the execution of $OptimalM()$. After the router is configured with these parameters and begins to measure the network traffic, it also monitors the value of n^*. If the maximum number of distinct contacts in a measurement period changes significantly, the values of m, s, p and T will be recomputed.

3.4 Experiments

3.4.1 Experimental Setup

We evaluate the performance of ESC and compare it with existing work, including the Two-level Filtering Algorithm (TFA) [14], the Thresholded Bitmap Algorithm (TBA) [4], and the Compact Spread Estimator (CSE) [15]. TFA uses two filters to reduce both the number of sources to be monitored and the number of contacts to be stored. It is designed to satisfy the classification objective in (3.1). TBA is not designed for meeting the classification objective. It cannot ensure that false positive/negative ratios are bounded. CSE is designed to estimate the spreads of external sources in a very compact memory space. It can be used for spreader classification by reporting the sources whose estimated spreads exceed a certain threshold. However, the design of CSE makes it unsuitable for meeting the objective in (3.1).

Online Streaming Module (OSM) [18] is another related work. We do not implement OSM in this study because Yoon et al. show that, given the same amount of

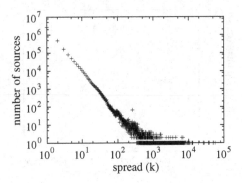

Fig. 3.4 Traffic distribution: each point shows the number of sources having a certain spread value

memory, CSE estimates spread values more accurately than OSM [15]. Moreover, the operations of OSM share certain similarity with Bloom filters. To store each contact, it performs three hash functions and makes three memory accesses. In comparison, ESC performs two hash functions and makes one memory access. Please refer to the related work section for more details on various spreader classification schemes.

The experiments use a real Internet traffic trace captured by Cisco's Netflow at the main gateway of our campus for a week. For example, in one day of the week, the traffic trace records 10,702,677 distinct contacts, 4,007,256 distinct source IP addresses and 56,167 distinct destination addresses. The average spread per source is 2.67, which means a source contacts 2.67 distinct destinations on average. Figure 3.4 shows the number of sources with respect to the source spread in log scale. The number of sources decreases exponentially as the spread value increases from 1 to 500. After that, there is zero, one or a few sources for each spread value.

We implement ESC, TFA, TBA and CSE, and execute them with the traffic trace as input. In the experiments, the source of a contact is the IP address of the sender and the destination is the IP address of the receiver. The measurement period is one day. The experimental results are the average over the week-long data.

One performance metric used in comparison is the amount of memory that is required for a spreader classification scheme to meet a given classification objective. Remarkably, the number of bits required by ESC is far smaller than the number of distinct sources in the traffic trace. That is, ESC requires much less than 1 bit per source to perform spreader classification. Other performance metrics include the false positive ratio and the false negative ratio, which will be explained further shortly.

3.4.2 Comparison in Terms of Memory Requirement

The first set of experiments compares ESC and TFA for the amount of memory that they need in order to satisfy a given classification objective, which is specified by four

Table 3.1 Memory requirements (in MB) of ESC, TFA and ESC-1 (i.e. ESC with $p = 1$) when $\alpha = 0.9$ and $\beta = 0.1$

h	$l = 0.1h$			$l = 0.3h$			$l = 0.5h$			$l = 0.7h$		
	ESC	TFA	ESC-1	ESC	TFA	ESC-1	ESC	TFA	ESC-1	ESC	TFA	ESC-1
500	0.09	2.02	0.33	0.19	2.53	0.43	0.30	3.61	0.54	0.97	6.12	1.01
1000	0.07	1.10	0.27	0.09	1.29	0.33	0.15	1.85	0.42	0.47	3.11	0.86
2000	0.03	0.55	0.24	0.05	0.71	0.29	0.08	1.02	0.42	0.25	1.62	0.86
3000	0.02	0.42	0.24	0.03	0.51	0.27	0.06	0.68	0.42	0.17	1.09	0.86
4000	0.01	0.32	0.21	0.03	0.38	0.27	0.03	0.52	0.42	0.13	0.83	0.86
5000	0.01	0.24	0.21	0.02	0.31	0.27	0.03	0.43	0.42	0.11	0.66	0.86

Table 3.2 Memory requirements (in MB) of ESC, TFA and ESC-1 (i.e. ESC with $p = 1$) when $\alpha = 0.95$ and $\beta = 0.05$

h	$l = 0.1h$			$l = 0.3h$			$l = 0.5h$			$l = 0.7h$		
	ESC	TFA	ESC-1	ESC	TFA	ESC-1	ESC	TFA	ESC-1	ESC	TFA	ESC-1
500	0.12	2.41	0.38	0.22	3.27	0.48	0.48	4.59	0.68	1.56	8.03	1.60
1000	0.08	1.29	0.32	0.12	1.65	0.38	0.24	2.34	0.50	0.76	4.04	1.20
2000	0.03	0.69	0.26	0.08	0.87	0.32	0.13	1.21	0.47	0.38	2.12	1.20
3000	0.02	0.46	0.26	0.06	0.60	0.32	0.09	0.83	0.47	0.26	1.42	1.20
4000	0.02	0.37	0.23	0.04	0.45	0.32	0.06	0.63	0.47	0.20	1.08	1.20
5000	0.01	0.29	0.23	0.04	0.35	0.32	0.05	0.52	0.47	0.16	0.89	1.20

parameters, α, β, h, and l. See Sect. 3.1 for the formal definition of the classification objective. We do not compare TBA and CSE here because they are not designed to meet this objective.

The memory required by ESC is determined based on the iterative algorithm in Sect. 3.3.3. The values of other parameters, s, T and p, are also decided by the same algorithm. For example, when $\alpha = 0.9$, $\beta = 0.1$, $h = 5000$, $l = 0.7h$, $n^* = 10M$, the system parameters are $s = 40$, $T = 4250$, $p = 0.01$, and $m = 0.11\,MB$. Using these parameters, we perform experiments on ESC with the traffic trace as input. The amount of memory required by TFA is determined experimentally based on the method in [14]. The parameters of TFA are chosen based on the original paper.

The memory requirements of ESC and TFA are presented in Tables 3.1, 3.2 and 3.3 with respect to α, β, h and l. For $\alpha = 0.9$ and $\beta = 0.1$, Table 3.1 shows that TFA requires 6–24 times of the memory that ESC requires, depending on the values of h and l (which the system administrator will select based on the organization's policy). For example, when $h = 500$ and $l = 0.5h$, ESC reduces the memory consumption by an order of magnitude when comparing with TFA.

Our experiments have also confirmed that the classification objective is indeed achieved by ESC. That is, the false positive ratio is always bounded by β and the false negative ratio is bounded by $1 - \alpha$ for all experiments reported in Tables 3.1, 3.2 and 3.3.

Table 3.3 Memory requirements (in MB) of ESC, TFA and ESC-1 (i.e. ESC with $p = 1$) when $\alpha = 0.99$ and $\beta = 0.01$

h	$l = 0.1h$			$l = 0.3h$			$l = 0.5h$			$l = 0.7h$		
	ESC	TFA	ESC-1	ESC	TFA	ESC-1	ESC	TFA	ESC-1	ESC	TFA	ESC-1
500	0.20	3.60	0.48	0.29	4.82	0.52	0.97	7.25	1.03	4.20	13.15	4.20
1000	0.10	1.92	0.38	0.15	2.42	0.40	0.50	3.60	0.67	1.59	6.54	3.10
2000	0.07	1.01	0.32	0.09	1.30	0.34	0.24	1.85	0.60	0.81	3.21	3.10
3000	0.04	0.68	0.29	0.07	0.85	0.34	0.16	1.24	0.60	0.53	2.18	3.10
4000	0.03	0.50	0.29	0.05	0.66	0.34	0.12	0.96	0.59	0.41	1.70	3.10
5000	0.03	0.42	0.29	0.05	0.55	0.34	0.10	0.77	0.59	0.33	1.38	3.10

To demonstrate the impact of probabilistic sampling, the table also includes the memory requirement of ESC when sampling is turned off (by setting $p = 1$). This version of ESC is denoted as ESC-1. Since p is set as a constant, the iterative algorithm in Sect. 3.3.3 needs to be slightly modified: the procedure $Optimal\,P\,(m, s)$ will always return 1, while other procedures remain the same. Table 3.1 shows that the memory saved by sampling is significant when h is large. For example, when $h = 5000$ and $l = 0.3h$, ESC with sampling uses less than one thirteenth of the memory that is needed by ESC-1. However, when h becomes smaller or $\frac{l}{h}$ becomes larger, ESC has to choose a larger sampling probability in order to limit the error in spread estimation caused by sampling. Consequently, it has to store more contacts and thus require more memory. For instance, when $h = 500$ and $l = 0.5h$, ESC with sampling uses 55.6 % of the memory that is needed by ESC-1.

Table 3.2 compares the memory requirements of ESC and TFA when $\alpha = 0.95$ and $\beta = 0.05$. Table 3.3 compares the memory requirements when $\alpha = 0.99$ and $\beta = 0.01$. They show similar results: (1) ESC uses significantly less memory than TFA, and (2) the probabilistic sampling method in ESC is critical for memory saving especially when h is large or $\frac{l}{h}$ is small. The tables also demonstrate that the memory requirement of either ESC or TFA increases when the classification objective becomes more stringent, i.e., α is set larger and β smaller.

TFA requires more memory because it stores the source and destination addresses of the contacts. In [17], the authors also indicate that Bloom Filters [2, 3] can be used to reduce the memory consumption. However, the paper does not give detailed design or parameter settings. Therefore, we cannot implement the Bloom-filter version of TFA. The paper claims that the memory requirement will be reduced by a factor of 2.5 when Bloom filters are used. Even when this factor is taken into account in Tables 3.1, 3.2 and 3.3, memory saving by ESC will still be significant.

Table 3.4 False negative ratio and false positive ratio of ESC, CSE and TBA with $m = 0.05$ MB

h	FNR			FPR		
	ESC	CSE	TBA	ESC	CSE	TBA
500	7.4e−2	0	2.6e−1	5.0e−2	1	9.0e−6
1000	1.0e−2	0	2.6e−1	5.5e−3	1	9.0e−6
2000	4.2e−3	0	2.5e−1	2.0e−3	1	1.1e−5
3000	5.5e−3	0	2.5e−1	2.0e−3	1	1.0e−5
4000	0	0	2.4e−1	2.0e−3	1	7.0e−6
5000	0	0	2.4e−1	2.0e−3	1	7.0e−6

Table 3.5 False negative ratio and false positive ratio of ESC, CSE and TBA with $m = 0.02$ MB

h	FNR			FPR		
	ESC	CSE	TBA	ESC	CSE	TBA
500	1.2e−2	3.3e−2	3.7e−3	1.5e−3	1.2e−1	1.8e−4
1000	8.8e−4	0	3.7e−3	7.5e−4	5.5e−2	1.9e−4
2000	0	0	9.3e−3	7.5e−4	5.5e−2	2.0e−4
3000	0	0	7.4e−3	7.5e−4	5.5e−2	1.8e−4
4000	0	0	1.9e−3	7.5e−4	5.5e−2	1.9e−4
5000	0	0	3.7e−3	7.5e−4	5.5e−2	1.8e−4

3.4.3 Comparison in Terms of False Positive Ratio and False Negative Ratio

The *false positive ratio* (FPR) is defined as the fraction of all non-heavy spreaders (whose spreads are no more than l) that are mistakenly reported as heavy spreaders. The *false negative ratio* (FNR) is the fraction of all heavy spreaders (whose spreads are no less than h) that are not reported by the system. In the previous section, we have shown that, given the bounds of FPR and FNR, it takes ESC much less memory to achieve the bounds than TFA. Since CSE and TBA are not designed for meeting a given set of bounds, we compare ESC with them by a different set of experiments that measure and compare the FPR and FNR values under a fixed amount of SRAM.

Given a fixed memory size m, we use $Optimal S(m, s)$ in Sect. 3.3.3 to determine the value of s in ESC, use $Optimal P(m, s)$ to determine the value of p, and then set the threshold T as $\frac{h+l}{2}$. We perform experiments using the week-long traffic trace. We average the daily FPR and FNR values over the week. For $m = 0.05$, and 0.2 MB, the results are presented in Tables 3.4 and 3.5, respectively. In both tables, $l = 0.5h$. We also perform the same experiments for CSE and TBA, and the results are presented in the tables as well. The optimal parameters are chosen for each scheme based on their original papers.

When the available memory is very small, such as 0.05 MB in Table 3.4, CSE has zero FNR but its FPR is 1.0, which means it reports all non-heavy spreaders. The reason is that, without probabilistic sampling, CSE stores information of too

Table 3.6 False negatives ratio and false positives ratio with $\alpha = 0.95$ and $\beta = 0.05$.

h	$l = 0.3h$		$l = 0.5h$		$l = 0.7h$	
	FNR	FPR	FNR	FPR	FNR	FPR
500	4.4e−3	3.6e−6	2.0e−3	1.0e−6	3.2e−4	0
1000	3.7e−3	3.0e−6	1.9e−3	1.0e−6	5.6e−4	0
2000	4.2e−3	1.0e−6	2.1e−3	1.0e−6	1.1e−3	1e−6
3000	1.8e−2	0	0	0	1.6e−3	0
4000	1.4e−2	1.0e−6	0	0	7.1e−3	0
5000	3.1e−3	0	6.4e−3	1.0e−6	0	1e−6

many contacts such that its data structure is fully saturated. In this case, the spread estimation method of CSE breaks down. TBA has a small FPR but its FNR is large. For example, when $h = 500$, its FNR is 26%. Only ESC achieves small values for both FNR and FPR. For example, when $h = 500$, its FNR is 7.4% and its FPR is 5.0%. These values decrease quickly as h increases. When $h = 1000$, they are 1.0 and 0.55%, respectively, while the FNR of TBA remains to be 26%. When the available memory increases in Table 3.5, the performance of all three schemes improves. Still, ESC performs better in most cases.

3.4.4 Performance of ESC

The iterative algorithm in Sect. 3.3.3 gives the memory requirement for meeting a certain classification objective in the worst case. In reality, the worst case scenario rarely happens. Hence, we expect the observed FPR to be much smaller than β and the observed FNR to be much smaller than $(1 - \alpha)$. This is indeed what we see in our experiments.

We run ESC on the week-long traffic trace under the following parameter settings: $\alpha = 0.95$, $\beta = 0.05$, h is varied from 500 to 5000, and l is varied from $0.3h$, $0.5h$ to $0.7h$. We use the iterative algorithm to determine the values of m and other system parameters. (Unlike the previous section, m is not a given value.) We then collect the daily values of FPR and FNR, and the average results are shown in Table 3.6. It shows that the real FPR/FNR are much smaller than the 5% objective (i.e., $1 - \alpha$ or β).

The reason is that the amount of memory m that ESC uses is determined based on the worst-case scenario, where the spreads of all non-heavy spreaders are l and the spreads of all heavy spreaders are h. Recall that in Sect. 3.3.2, we choose $k = l$ in (3.18) and $k = h$ in (3.17). However, in reality, not all non-heavy spreaders make the same number l of distinct contacts; many sources make very small numbers of contacts, as shown in Fig. 3.4. For a non-heavy spreader whose spread is much smaller than l, the probability for its estimated spread to exceed the threshold T (thus resulting in false positive) is certainly smaller than that of a non-heavy spreader whose

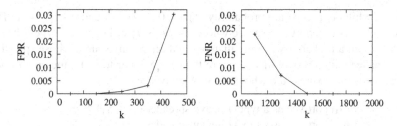

Fig. 3.5 False positive ratio and false negative ratio with $h = 1000$ and $l = 0.5h$

spread is l. Similarly, not all heavy spreaders make the same number h of distinct contacts. For a heavy spreader whose spread is much larger than h, the probability for its estimated spread to fall below the threshold (which results in false negative) is certainly smaller than that of a heavy spreader whose spread is h.

The above observation is also confirmed by experiment. We set $\alpha = 95\%$, $\beta = 5\%$, $h = 1000$, and $l = 0.5h$. After executing ESC on the traffic trace, we count the average daily number of false positives for sources whose spreads are in the range of $(0, l/5]$, $(l/5, 2l/5]$, ..., or $(4l/5, l]$, and the average daily number of false negatives for sources whose spreads are in the range of $[1000, 1,200)$, $[1200, 1400)$...The resulting FPR/FNR values in those ranges are presented in Fig. 3.5, where each point represents FPR (or FNR) for in a certain range of spread values (on the horizontal axis). The figure shows that FPR decreases quickly to zero for sources whose spreads are very small and FNR decreases quickly to zero for sources whose spreads are very large.

3.5 Multi-Objective Spreader Classification

So far we have considered spreader classification with a single objective: for example, reporting sources whose spreads are 100 or more with at least 95% probability, while reporting sources whose spreads are 75 or less with at most 5% probability. In practice, one may want to configure a router with more than one objective. For instance, in addition to the above objective of reporting *modestly heavy spreaders*, we may want to add another objective to identify more aggressive ones: reporting sources whose spreads are 1000 or more as *heavy spreaders* with at least 99% probability, and reporting sources whose spreads are 500 or less with at most 1% probability. How to efficiently perform spreader classification with multiple objectives is the subject of this section.

3.5.1 Problem Definition

Suppose there are t objectives. The jth objective, for any $j \in [1, t]$, is defined by four parameters, h_j, l_j, α_j and β_j. Among them, h_j and l_j are positive integers with

$h_j > l_j$, while α_j and β_j are probability values, $0 < \alpha_j < 1$ and $0 < \beta_j < 1$. The jth objective is to report any source whose spread is h_j or larger as a type-j heavy spreader with a probability no less than α_j, and report any source whose spread is l_j or smaller with a probability no more than β_j. Similar to (3.1), the conditional probabilities that express the whole set of objectives are

$$
\begin{aligned}
&Prob\{\text{report } src \text{ as a type-1 heavy spreader} \mid k \geq h_1\} \geq \alpha_1 \\
&Prob\{\text{report } src \text{ as a type-1 heavy spreader} \mid k \leq l_1\} \leq \beta_1 \\
&\qquad \cdots \\
&Prob\{\text{report } src \text{ as a type-j heavy spreader} \mid k \geq h_j\} \geq \alpha_j \\
&Prob\{\text{report } src \text{ as a type-j heavy spreader} \mid k \leq l_j\} \leq \beta_j \qquad\qquad (3.23)\\
&\qquad \cdots \\
&Prob\{\text{report } src \text{ as a type-t heavy spreader} \mid k \geq h_t\} \geq \alpha_t \\
&Prob\{\text{report } src \text{ as a type-t heavy spreader} \mid k \leq l_t\} \leq \beta_t,
\end{aligned}
$$

where src is an arbitrary source and its spread is k. Each objective is expressed by a pair of conditional probabilities. The goal is to minimize the amount of SRAM that is needed for achieving the above objectives.

3.5.2 Multi-Objective Spreader Classification Scheme and Optimal System Parameters

The spreader classification scheme in Sect. 3.2 can be extended to detect heavy spreaders under multiple objectives. The probabilistic sampling, dynamic bit sharing, and maximum likelihood-based spread estimation stay the same. The difference is how to determine the optimal system parameters. For spreader classification with one objective, we determine a threshold value T in Sect. 3.3 and report all sources whose estimated spreads exceed the threshold. The optimal values of T, s, p and m are computed based on (3.19). For spreader classification with multiple objectives, we need to determine one threshold for each objective. If the estimated spread of a source exceeds the threshold T_j for the jth objective, the source is reported as a type-j heavy spreader, where $1 \leq j \leq t$.

Following a mathematical process similar to Sect. 3.3, we can derive the set of constraints that the system parameters must satisfy in order to meet all objectives. We want to minimize the amount of memory that is needed to satisfy the constraints.

$$
Minimize\ m \qquad\qquad (3.24)
$$

$$
Subject\ to\ \forall 1 \leq j \leq t,
$$

$$
\sum_{i=0}^{\lfloor C_j \rfloor} \binom{s}{i} \cdot q(h_j)^i \cdot (1 - q(h_j))^{s-i} \geq \alpha_j, \qquad\qquad (3.25)
$$

$$\sum_{i=0}^{\lfloor C_j \rfloor} \binom{s}{i} \cdot q(l_j)^i \cdot (1 - q(l_j))^{s-i} \leq \beta_j, \tag{3.26}$$

$$\text{where } C_j = s \cdot \left(1 - \frac{p}{m}\right)^n \cdot \left(\frac{1 - \frac{p}{s}}{1 - \frac{p}{m}}\right)^{T_j}.$$

The conditional probabilities for the jth objective are transformed into two equivalent constraints: $Prob\{reports\text{rc} \mid k \geq h_j\} \geq \alpha_j$ is transformed to (3.25) and $Prob\{reports\text{rc} \mid k \leq l_j\} \leq \beta_j$ to (3.26). Our goal is to determine the optimal values for T_j ($1 \leq j \leq t$), s and p, such that they together minimize m. Note that there exists one threshold for each objective. In total, there are t thresholds. But s, p and m are common parameters shared for all objectives.

The algorithm in Sect. 3.3.3 can be extended to solve the above constrained optimization problem. We briefly describe the solution below. Consider the left side of (3.25) as a function $F_{h_j}(m, s, p, T_j)$ and the left side of (3.26) as a function $F_{l_j}(m, s, p, T_j)$. Namely, $\forall 1 \leq j \leq t$,

$$F_{h_j}(m, s, p, T_j) = \sum_{i=0}^{\lfloor C_j \rfloor} \binom{s}{i} \cdot q(h_j)^i \cdot (1 - q(h_j))^{s-i},$$

$$F_{l_j}(m, s, p, T_j) = \sum_{i=0}^{\lfloor C_j \rfloor} \binom{s}{i} \cdot q(l_j)^i \cdot (1 - q(l_j))^{s-i}.$$

We modify the iterative numerical algorithm in Sect. 3.3.3 to determine the optimal system parameters. The revised algorithm consists of five procedures, which are described as follows.

- First, we overload the procedure $Potential(m, s, p)$ in Sect. 3.3.3 and add one input parameter, j, indicating which functions the procedure is applied to. More specifically, $\forall j \in [1, t]$, the new procedure $Potential(m, s, p, j)$ is applied to functions $F_{h_j}(m, s, p, T_j)$ and $F_{l_j}(m, s, p, T_j)$. It uses the same binary search method as in Algorithm 1 to find the optimal value of T_j that maximizes the value of $F_{h_j}(m, s, p, T_j)$ under the condition that $F_{l_j}(m, s, p, T_j) \leq \beta$. The procedure returns the maximum value of $F_{h_j}(m, s, p, T_j)$, and as a byproduce, determines the optimal value of T_j. The pseudo code is the same as Algorithm 1 in the appendix except that $F_h(m, s, p, T)$ is replaced by $F_{h_j}(m, s, p, T_j)$ and $F_l(m, s, p, T)$ is replaced by $F_{l_j}(m, s, p, T_j)$.
- Second, we construct a new procedure called $PotentialAll(m, s, p)$, which takes a value of m, a value of s and a value of p as input and returns the minimum value of $\{Potential(m, s, p, 1) - \alpha_1, Potential(m, s, p, 2) - \alpha_2, .., Potential(m, s, p, t) - \alpha_t\}$. Clearly, if $PotentialAll(m, s, p) \geq 0$, all objectives can be satisfied. Essentially, the value of $PotentialAll(m, s, p)$ quantitatively indicates how conservative or aggressive we have chosen the value of m. If $PotentialAll(m, s, p)$ is positive, it means that the performance achieved by current memory size is more

than required. We shall reduce m. On the other hand, if $PotentialAll(m, s, p)$ is negative, we shall increase m.

- Third, given a value of m and a value of s, we construct a procedure $Optimal$ $P'(m, s)$ that determines the optimal value p^* such that $PotentialAll(m, s, p^*)$ is maximized. This procedure is similar to its counterpart in Sect. 3.3.3. The pseudo code is the same as Algorithm 2 in the appendix except that $Potential(m, s, p)$ is replaced with $PotentialAll(m, s, p)$.
- Fourth, given a value of m, we construct a procedure $OptimalS'(m)$ that determines the optimal value s^* such that $PotentialAll(m, s^*, OptimalP'(m, s^*))$ is maximized. This procedure is the same as its counterpart in Sect. 3.3.3 (Algorithm 3 in the appendix), except that $Potential(...)$ is replaced with $PotentialAll(...)$ and $OptimalP(m, s)$ is replaced with $OptimalP'(m, s)$.
- Fifth, we construct a procedure $OptimalM'()$ that determines the minimum memory requirement m^* such that $PotentialAll(m, OptimalS'(m), Optimal$ $P'(m, OptimalS'(m))) \geq 0$ is satisfied. Again, this procedure is similar as its counterpart in Sect. 3.3.3 (Algorithm 4 in the appendix). We skip the detailed description, which is virtually identical to the description in Sect. 3.3.3. In practice, given the objectives that are specified by $t, \alpha_j, \beta_j, h_j$ and $l_j, 1 \leq j \leq t$, a network administrator sets $m = OptimalM'(), s = OptimalS'(m)$, $p = OptimalP'(m, s)$, and $T_j(1 \leq j \leq t)$ as the threshold value before the last call of $Potential(m, s, p, j)$ is returned during the execution of $OptimalM'()$.

3.5.3 Additional Experimental Results

We perform additional experiments to evaluate our multi-objective spreader classification scheme. We use the same traffic trace as described in Sect. 3.4. We do not implement TFA, TBA or CSE because none of them can be applied for multi-objective spreader classification. In the new experiments, we let $t = 2$, i.e., there are 2 objectives, which are specified as follows: $h_1 = 500$, $\alpha_1 = 0.9$, $\beta_1 = 0.1$, $h_2 = 5000$, $\alpha_2 = 0.99$, and $\beta = 0.01$.

In the first set of experiments, we let $l_2 = 0.5h_2$ and vary l_1 from $0.1h_1$, $0.3h_1$, $0.5h_1$, to $0.7h_1$. We use the iterative algorithm in the previous section to compute the minimum amount of memory needed, as well as the optimal values for other system parameters. After that, we perform experiments based on these system parameters to report the heavy spreaders in the traffic trace. We measure the FPR and FNR values that are observed in the experiments. The results are presented in Table 3.7. The FNR values for type-1 heavy spreaders are shown in the column labeled with FNR1; they are indeed smaller than $1 - \alpha_1$. The FPR values for type-1 heavy spreaders are shown in the column labeled with FPR1; they are smaller than β_1 as required. Similarly, the data in columns FNR2 and FPR2 show that the second objective (specified by α_2 and β_2) is also met.

In the second set of experiments, we let $l_1 = 0.5h_1$ and vary l_2 from $0.1h_2$, $0.3h_2, 0.5h_2$, to $0.7h_2$. Again we use the iterative algorithm to determine the system

Table 3.7 Memory requirement, false negative ratio and false positive ratio with $h_1 = 500$, $h_2 = 5000$, and $l_2 = 0.5h_2$

l_1	m (MB)	$\alpha_1 = 0.9, \beta_1 = 0.1$		$\alpha_2 = 0.99, \beta_2 = 0.01$	
		FNR1	FPR1	FNR2	FPR2
$0.1h_1$	0.2	5.0e−3	1.7e−2	0	0
$0.3h_1$	0.3	8.0e−3	4.0e−4	0	0
$0.5h_1$	0.4	4.0e−3	1.0e−6	0	0
$0.7h_1$	1.1	2.0e−3	1.0e−6	0	0

Table 3.8 Memory requirement, false negative ratio and false positive ratio with $h_1 = 500$, $h_2 = 5000$, and $l_1 = 0.5h_1$

l_2	m (MB)	$\alpha_1 = 0.9, \beta_1 = 0.1$		$\alpha_2 = 0.99, \beta_2 = 0.01$	
		FNR1	FPR1	FNR2	FPR2
$0.1h_2$	0.3	1.5e−2	8.7e−5	0	0
$0.3h_2$	0.4	4.0e−3	2.0e−6	0	0
$0.5h_2$	0.4	4.0e−3	3.0e−6	0	3.0e−6
$0.7h_2$	0.6	5.0e−3	5.0e−6	0	1.1e−5

parameters and run experiments to measure the FPR and FNR values. The results are presented in Table 3.8. The data are interpreted in a similar way as we do for Table 3.7. Clearly, both objectives are met.

3.6 Other Methods

Venkataraman et al. [14] use hash tables to store the addresses of the sampled contacts. Their main contribution is to derive the optimal sampling probability that achieves a classification objective with pre-specified upper bounds on false-positive ratio and false-negative ratio. However, because their algorithms store the contact addresses, it leaves great room for improvement. Even if Bloom filters are used, the room for improvement is still significant, as we have argued in Sect. 3.4.

Estan et al. [10] propose a variety of bitmap algorithms to store the contacts (or active flows in their context). It saves space because each destination address is stored as a bit. However, assigning one bitmap to each source is not cheap if the average number of contacts per source is small. In addition, an index structure is needed to map a source to its bitmap. It is typically a hash table where each entry stores a source address and a pointer to the corresponding bitmap. Cao et al. [4] develop the thresholded bitmap algorithm based on the virtual bitmap algorithm presented in [10] for spread estimation. They use probabilistic sampling to reduce the information to be stored.

Zhao et al. [18] share a set of bitmaps among all sources. The scheme assigns three pseudo-randomly selected bitmaps to each source. When the source contacts a destination, the destination is stored by setting one bit in each of the three bitmaps. Because the bitmaps are shared by others, the information stored for one source becomes noise for others. Yoon et al. [15] observe that the noise introduced by sharing bitmaps cannot be appropriately removed if the number of bitmaps is not sufficiently large. By sharing bits instead of bitmaps, CSE considerably reduces the memory consumption.

Also related is the work by Bandi et al. [1] on the heavy distinct hitter problem, which is essentially the same as spreader classification. Their algorithm exploits TCAM (Ternary Content Addressable Memory), a special kind of memory found in NPUs (Network Processing Units). The emphasis of their work is on the processing time.

A related branch of research is the detection of heavy-hitters [5–9, 11, 13, 16]. A heavy-hitter is a source that sends a lot of packets during a measurement period no matter whether the packets are sent to a few or many distinct destinations.

3.7 Summary

Spreader classification is an important network measurement function. The recent research trend is to implement such a function in the tight SRAM space to catch up with the rapid advance in network speed. This chapter presents an efficient spreader classification scheme based on a new method called *dynamic bit sharing*, which optimally combines probabilistic sampling, bit-sharing storage, and maximum likelihood estimation. We demonstrate theoretically and experimentally that this scheme is able to achieve a classification objective with arbitrarily-set bounds on worst-case false positive/negative ratios. It does so in a very tight memory space where the number of bits available is much smaller than the number of external sources to be monitored. In addition, the scheme can be extended to solve the multi-objective spreader classification problem.

Appendix: Algorithms for Optimal System Parameters

Algorithm 1 $Potential(m, s, p)$

INPUT: m, s, p and β

OUTPUT: The maximum value of $F_h(m, s, p, T)$ under the condition that $F_l(m, s, p, T) \leq \beta$.

Pick a small integer T_1 such that $F_l(m, s, p, T_1) > \beta$ and a large integer T_2 such that $F_l(m, s, p, T_2) \leq \beta$.

while $T_2 - T_1 > 1$ **do**

  $\bar{T} = \lfloor (T_1 + T_2)/2 \rfloor$

  if $F_l(m, s, p, \bar{T}) \leq \beta$ **then**

    $T_1 = \bar{T}$

  else

    $T_2 = \bar{T}$

  end if

end while

$T^* = \bar{T}$

return $F_h(m, s, p, T^*)$

Algorithm 2 $OptimalP(m, s)$

INPUT: m, s and δ

OUTPUT: The optimal value of p^* such that $Potential(m, s, p^*)$ is maximized

---

$p_1 = 0, p_2 = 1$

while $p_2 - p_1 > \delta$ **do**

  $\bar{p} = (p_1 + p_2)/2$

  if $Potential(m, s, \bar{p}) < Potential(m, s, \bar{p} + \delta)$ **then**

    $p_1 = \bar{p}$

  else

    $p_2 = \bar{p}$

  end if

end while

$p^* = (p_1 + p_2)/2$

return p^*

Algorithm 3 $OptimalS(m)$

INPUT: m
OUTPUT: The optimal value of s^* such that $Potential(m, s^*, OptimalP(m, s^*))$ is maximized

$s_1 = 1, s_2 = m$
while $s_2 - s_1 > 1$ **do**
$\quad \bar{s} = \lfloor (s_1 + s_2)/2 \rfloor$
$\quad$ **if** $Potential(m, \bar{s}, OptimalP(m, \bar{s})) < Potential(m, \bar{s} + 1, OptimalP(m, \bar{s} + 1))$ **then**
$\quad\quad s_1 = \bar{s}$
$\quad$ **else**
$\quad\quad s_2 = \bar{s}$
$\quad$ **end if**
end while
$s^* = \bar{s}$
return s^*

Algorithm 4 $OptimalM()$

OUTPUT: The smallest value m^* that satisfies $Potential(m^*, ...) \geq \alpha$

Pick a small value m_1 such that $Potential(m_1, ...) \leq \alpha$ and a large value m_2 such that $Potential(m_2, ...) \geq \alpha$.
while $m_2 - m_1 > 0$ **do**
$\quad \bar{m} = \lfloor (m_1 + m_2)/2 \rfloor$
$\quad$ **if** $Potential(\bar{m}, ...) \leq \alpha$ **then**
$\quad\quad m_1 = \bar{m}$
$\quad$ **else**
$\quad\quad m_2 = \bar{m}$
$\quad$ **end if**
end while
$m^* = \bar{m}$
return m^*

References

1. Bandi, N., Agrawal, D., Abbadi, A.: Fast algorithms for heavy distinct hitters using associative memories. In: Proceedings of IEEE International Conference on, Distributed Computing Systems(ICDCS) (2007)
2. Bloom, B.H.: Space/time trade-offs in hash coding with allowable errors. Commun. ACM **13**(7), 422–426 (1970)
3. Broder, A., Mitzenmacher, M.: Network applications of bloom filters: a survey. Internet Math. **1**(4), 485–509 (2002)
4. Cao, J., Jin, Y., Chen, A., Bu, T., Zhang, Z.: Identifying high cardinality internet hosts. In: Proceedings of IEEE INFOCOM (2009)
5. Charikar, M., Chen, K., Farach-Colton, M.: Finding frequent items in data streams. In: Proceedings of International Colloquium on Automata, Languages, and Programming (ICALP) (2002)
6. Cormode, G., Muthukrishnan, S.: Space efficient mining of multigraph streams. In: Proceedings of ACM PODS (2005)
7. Demaine, E., Lopez-Ortiz, A., Ian-Munro, J.: Frequency estimation of internet pacet streams with limited space. In: Proceedings of Annual European Symposium on Algorithms (ESA) (2002)
8. Dimitropoulos, X., Hurley, P., Kind, A.: Probabilistic lossy counting: an efficient algorithm for finding heavy hitters. ACM SIGCOMM Comput. Commun. Rev. **38**(1), 7–16 (2008)
9. Estan, C., Varghese, G.: New directions in traffic measurement and accounting. In: Proceedings of ACM SIGCOMM (2002)
10. Estan, C., Varghese, G., Fish, M.: Bitmap algorithms for counting active flows on high-speed links. IEEE/ACM Trans. Netw. **14**(5), 925–937 (2006)
11. Gibbons, P., Matias, Y.: New sampling-based summary statistics for improving approximate query answers. In: Proceedings of ACM SIGMOD (1998)
12. Hwang, K., Vander-Zanden, B., Taylor, H.: A linear-time probabilistic counting algorithm for database applications. ACM Trans. Database Syst. **15**(2), 208–229 (1990)
13. Manku, G., Motwani, R.: Approximate frequency counts over data streams. In: Proceedings of VLDB (2002)
14. Venkatataman, S., Song, D., Gibbons, P., Blum, A.: New streaming algorithms for fast detection of superspreaders. In: Proceedings of NDSS (2005)
15. Yoon, M., Li, T., Chen, S., Peir, J.: Fit a spread estimator in small memory. In: Proceedings of IEEE INFOCOM (2009)
16. Zhang, Y., Singh, S., Sen, S., Duffield, N., Lund, C.: Online identification of hierarchical heavy hitters: algorithms, evaluation, and application. In: Proceedings of ACM SIGCOMM IMC (2004)
17. Zhao, Q., Kumar, A., Xu, J.: Joint Data Streaming and Sampling Techniques for Detection of Super Sources and Destinations. Proc. of USENIX/ACM Internet Measurement Conference (2005).
18. Zhao, Q., Xu, J., Kumar, A.: Detection of Super Sources and Destinations in High-Speed Networks: Algorithms, Analysis and Evaluation. IEEE J. on Selected Areas in Communications (JASC) 24(10), 1840–1852 (2006).

Chapter 4
Origin–Destination Flow Measurement

Abstract This chapter presents an efficient approach for origin–destination flow measurement in high-speed networks, where an origin–destination (OD) flow between two routers is the set of packets that pass both routers. The OD flow measurement has wide usage in many network management applications. We consider two performance metrics, measurement efficiency and accuracy. The former requires measurement functions to minimize per-packet processing overhead in order to keep up with the line speeds of today's high-speed networks. The latter requires measurement functions to achieve accurate measurement results with small bias and standard deviation. We present a novel measurement method that employs a compact data structure for packet information storage and uses a new statistical inference approach for OD flow measurement. Both simulations and experiments are performed to demonstrate the effectiveness of our method. The rest of this chapter is organized as follows: Section 4.1 gives the problem statement and performance metrics. Section 4.2 presents a novel origin-destination flow measurement method. Section 4.3 discusses the simulation results. Section 4.4 presents the experimental results. Section 4.5 describes other related methods. Section 4.6 gives the summary.

Keywords Origin–destination flow estimator · Bitmap

4.1 Problem Statement and Performance Metrics

4.1.1 Problem Statement

Let S be a subset of routers of interest in a network. The problem is to measure traffic volume between any pair of routers in S. We model an origin-destination (OD) flow as the set of packets traverse between two routers (the undirectional case) or traverse from one router to the other (the directional case). Our goal is to measure the size of each OD flow in terms of number of packets.

T. Li and S. Chen, *Traffic Measurement on the Internet*,
SpringerBriefs in Computer Science, DOI: 10.1007/978-1-4614-4851-8_4,
© The Author(s) 2012

Consider the set of access routers on the perimeter of an ISP network. If each access router stores information about ingress packets (that enter the ISP network) and egress packets (that leave the ISP network) in separate data structures, we can figure out the size of a directional OD flow by comparing the information in the ingress data structure of the origin router and the information in the egress data structure of the destination router. On the other hand, if each access router stores information of all arrival packets in the same data structure, we can figure out the size of an undirectional OD flow by comparing the information in the data structures of both routers. The measurement method presented in this chapter can be applied to both cases even though our description uses the undirectional case for simplicity.

We consider two performance metrics, per-packet processing overhead and measurement accuracy, which are discussed below.

4.1.2 Per-Packet Processing Overhead

The maximum packet throughput that an online measurement function can achieve is determined by the per-packet processing overhead of the function. In order to keep up with today's high-speed network, it is desirable to make the per-packet processing overhead as small as possible, especially when the SRAM and processing circuits are shared by other critical functions for routing, packet scheduling, traffic management and security purposes.

The per-packet processing overhead is mainly determined by the computational complexity and the number of memory accesses for each packet. When a router receives a packet, it needs to perform certain computation to determine the proper location for the information storage and at least one memory access for the storing operation. We will show that our OD flow measurement function is able to achieve extremely small per-packet processing overhead.

4.1.3 Measurement Accuracy

Let n_c be the true size of an OD flow size between two routers and $\hat{n}_c$ be the measured value. The accuracy requirement is given as follows: the probability for n_c to fall in the interval $[\hat{n}_c \cdot (1 - \beta), \hat{n}_c \cdot (1 + \beta)]$ must be at least α, where α and β are system parameters in the range of $(0, 1)$. For example, when $\alpha = 95\%$ and $\beta = 0.1$, it means that the measured size has a probability of 95 % to be within $\pm 10\%$ of the true value.

4.2 Origin–Destination Flow Measurement

We first describe two straightforward approaches and discuss their limitations. We then motivate the bitmap idea that we use in this study. Finally we present a novel origin–destination flow measurement method (ODFM) in details.

4.2.1 Straightforward Approaches and Their Limitations

A straightforward approach is for each router to store the information of all packets that pass it. In this way, when we want to measure the OD flow size of two routers, we only need to compare the two sets of packet information and count how many packets the two sets have in common, i.e., the cardinality of the intersection of the two sets. Clearly, storing information of all packets is unrealistic since the number of packets passing a router is huge in high-speed networks and it imposes an extremely large memory requirement on the router.

In order to reduce the memory requirement, we can store the signatures of packets instead. The signature of a packet is a hash value of the packet with a fixed length. When the length of the signature is long enough, e.g., 160 bits if using SHA-1 [21], the chance of two packets having the same signatures is negligibly small. Therefore, we can count the number of identical signatures that stored in the two routers to obtain the OD flow size. This enhancement can reduce the memory requirement to some extent. However, it is still not memory efficient. Suppose there are 1 M packets that pass a router during a measurement period. When the length of the signature is 160 bits long, a router needs 20 MB ($1M \times 160/8$) memory to store the information of all signatures, which is still too much in practise. Using smaller signatures cannot solve the problem. For example, if we reduce the signature length to just 16 bits, the memory requirement is still 2 MB. We want to control space overhead to less than 1 bit per packet.

Another approach is for a router to maintain counters for all other routers in the network. A packet keeps track of the routers that it has traversed in its header. When a router receives a packet, it first checks the packet header and knows which routers this packet has passed. It then increases the corresponding counters by one. Then the router adds its own address into the packet before sending it out. At the end of a measurement period, in order to obtain the OD flow size of router r_1 and router r_2, we first check r_1 and find the counter for r_2, which stores the number of packets that enter r_2 and exit from r_1. We then check r_2 and find the counter for r_1, which stores the number of packets that enter r_1 and exit from r_2. The summation of the two counters is the OD flow size of r_1 and r_2. Although the computation for the OD flow size is very simple at the end of the measurement period, this approach has two main drawbacks during the packet processing period. First, the router needs to extract the addresses of other routers from the packet header and may have to update multiple counters. It also needs to insert its own address to the packet header and may have to recompute the checksum field. All these will slow down packet forwarding. Second, the packet size changes after each router, which may result in frequent fragmentation.

4.2.2 ODFM: Motivation and Overview

We design a bitmap based OD flow measurement method that is able to solve the problems that the above two approaches have. Instead of storing the signatures of packets, each router maintains a bit array (called *bitmap*) of a fixed length. At the beginning of a measurement period, all bits are set to zeros. We require routers to implement a common hash function. When a router receives a packet, it pseudo-randomly maps the packet to one bit in the array using the common hash function and then sets the bit to one. At the end of the measurement period, we measure the OD flow size of two routers by comparing their bit arrays. Clearly, since the same hash function is used, a packet will always *choose* (i.e., be mapped to) the same bit location in the arrays of all routers that it traverses. Therefore, if a packet enters router r_1 and exits from r_2 or the other way around, its corresponding bit in these two bit arrays must be both set to one. Based on this observation, we can take a bitwise AND operation of the two bit arrays and count the number of ones in the combined bit array to measure the OD flow.

A closer look may suggest that this approach can potentially result in overestimation. Suppose two packets, called p_1 and p_2, map to the same bit location j by the hash function. While p_1 passes one router and p_2 passes another. In this case, the jth bit of both arrays at the two routers will be set to one. When we compare the two bit arrays, we will falsely treat p_1 and p_2 as the same packet and overestimate the OD flow size. However, there is a nice property of our scheme: because the bit for each packet is pseudo-randomly picked from the array, any two packets has equal probability to choose a common bit. When the number of packets and the size of the bit array are large enough, this event occurs in the bit array uniformly at random and the overestimation problem can be removed through statistical analysis. This property enables us to design a compact yet accurate measurement method.

4.2.3 ODFM: Storing the Packet Information

ODFM consists of two components: one for storing the packet information into routers, the other for measuring the OD flow of any two routers. Below we presents the first component. At the beginning of a measurement period, each router maintains a bit array B with a fixed length m. Initially each bit in B is set to zero. The ith bit in the array is denoted as $B[i]$. When a router receives a packet p, it pseudo-randomly picks one bit in B by performing a hash operation $H(p)$ and set the bit to one, where $H(..)$ is a hash function whose output range is $[0..m-1]$. More specifically, to store

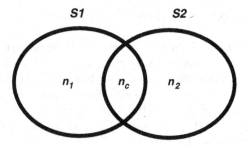

Fig. 4.1 The relation between two routers r_1 and r_2

the packet p, ODFM performs the following assignment:

$$B[H(p)] := 1. \tag{4.1}$$

It is worth noting that a router only needs to perform one hash operation and sets one bit in its bit array per packet. The hash operation does not have to be performed on the entire packet. Instead, it is applied only to the invariant fields in the header, such as source IP address, destination IP address, etc. For two packets that are fragments of the same original packet, although they share the same source/destination addresses as well as identification number, their fragmentation offset values are different.

4.2.4 ODFM: Measuring the Size of Each OD Flow

At the end of a measurement period, all routers will report their bit arrays to a centralized server, e.g., the network management center, which performs the offline measurement. ODFM employs the maximum likelihood estimation (MLE) [6] to measure the OD flow of any two routers based on their bit arrays. Let S_1 and S_2 be the sets of packets that pass routers r_1 and r_2, respectively. Let n_1 and n_2 be the cardinalities of S_1 and S_2, i.e., $n_1 = |S_1|$ and $n_2 = |S_2|$. Let n_c be the number of common packets that r_1 and r_2 share, i.e., the OD flow size of the two routers, which is the value that we want to measure. Figure 4.1 illustrates the relationship of n_1, n_2 and n_c, where $n_c = |S_1 \cap S_2|$. Let B_1 and B_2 be the bit arrays of r_1 and r_2, respectively. Let U_1 and U_2 be the numbers of '0's in B_1 and B_2, respectively. Let V_1 and V_2 be the percentages of bits in B_1 and B_2 whose values are zero. Namely, $V_1 = \frac{U_1}{m}$ and $V_2 = \frac{U_2}{m}$.

The measurement process consists of two steps. In the first step, we compute the cardinality of S_1 (i.e., n_1) and the cardinality of S_2 (i.e., n_2) from B_1 and B_2 based on the probabilistic counting method in [12]. A simpler approach is for a router to use a counter to keep track of the number of packets that it receives in a measurement period. The counter may be implemented by a register on the processor.

The second step measures the value of n_c. We take a bitwise AND operation of B_1 and B_2. The result is denoted as B_c. Namely,

$$B_c[i] = B_1[i] \& B_2[i], \forall i \in [0..m-1]. \tag{4.2}$$

For an arbitrary bit b in B_c, it is '0' if and only if the following two conditions are both satisfied: *first, it is not chosen by any packet in $S_1 \cap S_2$.* If b is chosen by a packet $p \in S_1 \cap S_2$, we know the corresponding bits in both B_1 and B_2 will be set to '1'. Therefore, b will be '1'. *Second, it is either not chosen by any packet in $S_1 - S_2$ or not chosen by any packet $S_2 - S_1$.* If it is chosen by both a packet $p_1 \in S_1 - S_2$ and a packet $p_2 \in S_2 - S_1$, the corresponding bits in both B_1 and B_2 will be also set to '1'. As a result, b will be '1'.

For the first condition, each packet in $S_1 \cap S_2$ has probability $\frac{1}{m}$ to set b to '1', which means the probability for b not to be set by this packet is $1 - \frac{1}{m}$. There are n_c packets in $S_1 \cap S_2$. Therefore, the probability for b not to be set to '1' by any packet in $S_1 \cap S_2$ is $(1 - \frac{1}{m})^{n_c}$. For the second condition, the probability for b not to be chosen by any packet in $S_1 - S_2$ is $(1 - \frac{1}{m})^{n_1 - n_c}$ and the probability for it not to be chosen by any packet in $S_2 - S_1$ is $(1 - \frac{1}{m})^{n_2 - n_c}$. Combining the above analysis, the probability $q(n_c)$ for b to remain '0' in B_c is

$$q(n_c) = (1 - \frac{1}{m})^{n_c} \{ 1 - (1 - (1 - \frac{1}{m})^{n_1 - n_c}) $$
$$\times (1 - (1 - \frac{1}{m})^{n_2 - n_c}) \}$$
$$= (1 - \frac{1}{m})^{n_1} + (1 - \frac{1}{m})^{n_2} - (1 - \frac{1}{m})^{n_1 + n_2 - n_c}. \tag{4.3}$$

Each bit in B_c has a probability $q(n_c)$ to be '0'. The observed number of '0' bits in B_c is U_c. Therefore, the likelihood function for this observation to occur is given as follows:

$$L = q(n_c)^{U_c} \times (1 - q(n_c))^{m - U_c}. \tag{4.4}$$

Next, we follow the standard process of maximum likelihood estimation to find the optimal value of n_c that maximizes the above likelihood function:

$$\hat{n}_c = \arg \max_{n_c} \{L\}. \tag{4.5}$$

To find $\hat{n}_c$, we take a logarithm operation to both sides of (4.4).

$$\ln L = U_c \times \ln q(n_c) + (m - U_c) \times \ln(1 - q(n_c)). \tag{4.6}$$

We then differentiate the above equation:

$$\frac{d \ln L}{d n_c} = (\frac{U_c}{q(n_c)} - \frac{m - U_c}{1 - q(n_c)}) \times q'(n_c)$$

$$= (\frac{U_c}{q(n_c)} - \frac{m - U_c}{1 - q(n_c)}) \times \ln(1 - \frac{1}{m})$$

$$\times (1 - \frac{1}{m})^{n_1 + n_2 - n_c}. \qquad (4.7)$$

From (4.3), we have

$$q'(n_c) = \frac{dq(n_c)}{dn_c}$$

$$= \ln(1 - \frac{1}{m}) \times (1 - \frac{1}{m})^{n_1 + n_2 - n_c}. \qquad (4.8)$$

In order to compute $\hat{n}_c$, we set the right side of (4.7) to zero, i.e.

$$(\frac{U_c}{q(n_c)} - \frac{m - U_c}{1 - q(n_c)}) \times \ln(1 - \frac{1}{m}) \times (1 - \frac{1}{m})^{n_1 + n_2 - n_c} = 0 \qquad (4.9)$$

Since neither $\ln(1 - \frac{1}{m})$ nor $(1 - \frac{1}{m})^{n_1 + n_2 - n_c}$ can be 0 when m is positive, we have

$$\frac{U_c}{q(n_c)} - \frac{m - U_c}{1 - q(n_c)} = 0. \qquad (4.10)$$

Applying (4.3) to (4.10), we have

$$(1 - \frac{1}{m})^{n_1} + (1 - \frac{1}{m})^{n_2} - (1 - \frac{1}{m})^{n_1 + n_2 - n_c} = \frac{U_c}{m}$$

$$= V_c. \qquad (4.11)$$

In above equation, m, n_1, and n_2 are all known values, and V_c can also be computed when the packets information are recorded. As a result, we can measure n_c in the following formula:

$$n_c = n_1 + n_2 - \frac{\ln((1 - \frac{1}{m})^{n_1} + (1 - \frac{1}{m})^{n_2} - V_c)}{\ln(1 - \frac{1}{m})}. \qquad (4.12)$$

4.2.5 Measurement Accuracy

We analyze the measurement accuracy. According to the standard theory of MLE [16], when the values of m, n_1, and n_2 are large enough, the measured OD flow size $\hat{n}_c$ approximately follows a normal distribution:

$$\hat{n}_c \sim Norm\left(n_c, \frac{1}{\mathscr{I}(\hat{n}_c)}\right), \tag{4.13}$$

where $\mathscr{I}(\hat{n}_c)$ is the fisher information[1] of L, which is defined as follows

$$\mathscr{I}(\hat{n}_c) = -E\left[\frac{d^2\ln L}{dn_c^2}\right]. \tag{4.14}$$

According to (4.7), we compute the second-order derivative of $\ln L$

$$\begin{aligned}
\frac{d^2\ln L}{dn_c^2} &= \ln(1 - \frac{1}{m}) \times \left[\left(-\frac{U_c \cdot q'(n_c)}{q^2(n_c)} - \frac{(m - U_c) \cdot q'(n_c)}{(1 - q(n_c))^2}\right)\right. \\
&\quad \left. \times C - \left(\frac{U_c}{q(n_c)} - \frac{m - U_c}{1 - q(n_c)}\right) \times C\right],
\end{aligned} \tag{4.15}$$

where $C = (1 - \frac{1}{m})^{n_1+n_2-n_c}$ and $q'(n_c)$ is given in (4.8).

We use the probabilistic counting method [12] to compute the expected value of U_c. Let X_i be the event that the ith bit in B_c remains '0' at the end of the measurement period and 1_{X_i} be the corresponding indicator random variable. As the size of B_c is m, for an arbitrary bit b, it has probability $q(n_c)$ to remain '0'. U_c is the number of '0's in B_c, $U_c = \sum_{i=0}^{m-1} 1_{X_i}$. Hence,

$$E(U_c) = \sum_{i=0}^{m-1} E(1_{X_i}) = \sum_{i=0}^{m-1} q(n_c) = m \cdot q(n_c) \tag{4.16}$$

Therefore, we have

$$\begin{aligned}
\mathscr{I}(\hat{n}_c) &= -E\left[\frac{d^2\ln L}{dn_c^2}\right] \\
&= \ln(1 - \frac{1}{m}) \times \left(\frac{m \cdot q'(n_c)}{q(n_c)} + \frac{m \cdot q'(n_c)}{1 - q(n_c)}\right) \times C, \tag{4.17}
\end{aligned}$$

as the expected value of $(\frac{U_c}{q(n_c)} - \frac{m-U_c}{1-q(n_c)})$ is 0.

According to (4.13), the variance of $\hat{n}_c$ is

---

[1] The fisher information [13] is a way of measuring the amount of information that an observable random variable x carries about an unknown parameter θ upon which the likelihood function of θ, $L(\theta) = f(x; \theta)$, depends.

$$Var(\hat{n}_c) = \frac{1}{\mathscr{I}(\hat{n}_c)}$$

$$= \frac{1}{\ln(1 - \frac{1}{m}) \times \left(\frac{m \cdot q'(n_c)}{q(n_c)} + \frac{m \cdot q'(n_c)}{1 - q(n_c)} \right) \times C}. \qquad (4.18)$$

and the confidence interval of our measurement is

$$\hat{n}_c \pm \frac{Z_\alpha}{\sqrt{\ln(1 - \frac{1}{m}) \times \left(\frac{m \cdot q'(n_c)}{q(n_c)} + \frac{m \cdot q'(n_c)}{1 - q(n_c)} \right) \times C}}, \qquad (4.19)$$

where α is the confidence level parameter and Z_α is the α percentile for the standard Gaussian distribution [3]. For example, when $\alpha = 99\%$, $Z_\alpha = 2.58$.

4.3 Simulations

We evaluate the performance of the method ODFM by simulations in this section. We will present experimental results based on real traffic trace in the next section. In both simulations and experiments, we compare ODFM with the most related work, QMLE [4]. For fair comparison, we assign the same amount of memory to ODFM and QMLE. We compare them in terms of online processing overhead and measurement accuracy.

Simulations are performed under system parameters, n_1, n_2, and n_c. For an origin–destination router pair, n_1 is the number of packets that one router receives during the measurement period, and n_2 is the number of packets that the othe router receives. Parameter n_c is the actual OD flow size. The amount of memory used is set to be 1 MB.

In the first set of simulations, we let $n_1 = 6,000,000$, $n_2 = 6,000,000$, 300,000, or 100,000. We vary n_c from 100 to 50,000. We use ODFM and QMLE to measure the flow size, and compare it with n_c to see how accurate the measurement is.

In the second set of simulations, we model a more realistic scenario, where n_1, n_2 and n_c are randomly chosen. The values of n_1 and n_2 are randomly selected from the range of [100,000, 10,000,000], and the value of n_c is randomly selected from [100,50,000] in each simulation run.

4.3.1 Processing Overhead

Per-packet processing overhead of a measurement method is mainly determined by the number of memory accesses and the number of hash operations for each packet.

Table 4.1 Number of memory accesses and number of hash operations per packet with $n_1 = 6,000,000$ and $n_2 = 6,000,000$

	Memory accesses	Hash operations	Constant?
ODFM	1	1	Yes
QMLE	1.50	2	No

Table 4.2 Number of memory accesses and number of hash operations per packet with $n_1 = 6,000,000$ and $n_2 = 300,000$

	Memory accesses	Hash operations	Constant?
ODFM	1	1	Yes
QMLE	1.56	2	No

Table 4.3 Number of memory accesses and number of hash operations per packet with $n_1 = 6,000,000$ and $n_2 = 100,000$

	Memory accesses	Hash operations	Constant?
ODFM	1	1	Yes
QMLE	1.54	2	No

Table 4.4 Number of memory accesses and number of hash operations per packet with the values of n_1 and n_2 are randomly assigned between 100,000 and 10,000,000

	Memory accesses	Hash operations	Constant?
ODFM	1	1	Yes
QMLE	1.22	2	No

Table 4.1 shows the averaged results when $n_1 = 6,000,000$, $n_2 = 6,000,000$, and n_c varies from 100 to 50,000. ODFM requires only 1 hash operation and 1 memory access (memory write) for each packet, which is the optimal. QMLE requires more per-packet processing overhead. It incurs 1.50 memory accesses and 2 hash operations on average. Furthermore, per-packet processing overhead of ODFM is constant, while QMLE requires variable per-packet processing overhead, which is undesirable in practice. Table 4.2 and Table 4.3 present similar results with $n_2 = 300,000$ and $100,000$ respectively. Table 4.4 shows the results when the values of n_1 and n_2 are randomly chosen in the range $[100,000, 10,000,000]$ and the value of n_c is randomly chosen in the range of $[100, 50,000]$.

4.3.2 Measurement Accuracy

Figures 4.2, 4.4, 4.5 present the measurement results of ODFM and QMLE. Each figure consists of four plots. Each point in the first plot (ODFM) or the second plot

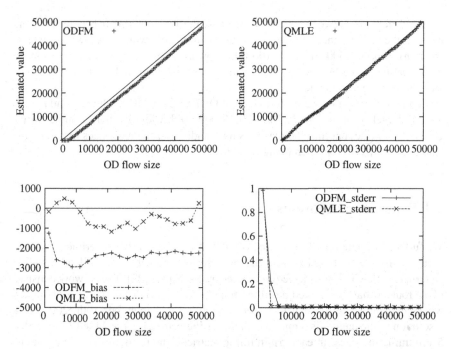

Fig. 4.2 *First plot (top left)*estimation results by ODFM when $n_1 = 6,000,000$ and $n_2 = 6,000,000$.*Second plot (top right)* estimation results by QMLE when $n_1 = 6,000,000$ and $n_2 = 6,000,000$. *Third plot (bottom left)* bias of ODFM and QMLE, which is the measured $E(\hat{n}_c - n_c)$ with respect to n_c. *Fourth plot (bottom right)* standard deviation of ODFM and QMLE, which is the measured $\frac{\sqrt{Var(\hat{n}_c)}}{n_c}$

(QMLE) represents an OD flow. The x-axis is the actual flow size n_c, and the y-axis is the estimated value $\hat{n}_c$. We also show the equality line, $y = x$, for reference. Clearly, the closer a point is to the equality line, the better the estimation result is. The third plot shows the corresponding measured bias of the first two plots, which is $E(\hat{n}_c - n_c)$. The fourth plot shows the corresponding standard deviation of the first two plots, which is $\frac{\sqrt{Var(\hat{n}_c)}}{n_c}$. In order to clearly present the estimation results of the two methods, we divide the horizontal coordinate into 25 measurement bins of width 2,000, and numerically measure the bias and standard deviation in each bin. The three figures present the following results.

As shown in the first plot of Fig. 4.2, when the values of n_1 and n_2 are the same, ODFM has a small bias in its measurement, which is understandable because it is well known that the maximum likelihood estimation may produce small bias under certain parameter settings. The second plot shows that QMLE performs better and produces almost perfect results. However, this is only part of the story. When the values of n_1 and n_2 are different, as shown in Fig 4.3 where $n_1 = 6,000,000$ and $n_2 = 300,000$, ODFM performs nearly perfectly, while QMLE produces large bias.

As the difference between n_1 and n_2 widens, the bias of QMLE becomes larger, whereas the performance of ODFM is actually improved, which is shown in Fig. 4.4 where $n_1 = 6,000,000$ and $n_2 = 100,000$. Now the question is which case is closer to the reality, n_1 and n_2 having close values or diverse values? It is the latter, as we will show in the next section.

Figure 4.5 compares the performance of ODFM and QDFM when n_1 and n_2 are randomly picked in the range $[100,000, 10,000,000]$. Clearly, ODFM outperforms QMLE by a wide margin. The reason is that randomly-selected values of n_1 and n_2 tend to be very different than being close to each other.

4.4 Experimental Results

We further evaluate the performance of ODFM and QMLE by experiments in this section. The experimental dataset that we use is obtained from Abilene network (Internet2) [2], which is collected and shared by Yin Zhang [22]. The network consists of 12 routers that are located at different cities in US [1]. The dataset contains 24 weeks of Abilene traffic matrices from March 1st to September 10th, 2004. The resolution of the dataset is 5 min, which means there are $24 \times 7 \times 24 \times 12 = 48,384$ 5 min traffic matrices. In each 5 min traffic matrices, the traffic flows of the routers range from 0.5 to 20 GB. We set the duration of a measurement period to 5 min and assume that the packet size is 1,500 bytes, which means the routers receive about 0.3 to 13M packets in one measurement period.

We allocate 1 MB memory resource to each router and implement the two measurement methods based on the 24 weeks' traffic matrices. The experimental results are similar for those weeks. In this section, we only present the results for the first week.

4.4.1 Number of Packets for an Origin–Destination Pair

Before measuring the size of each OD flow, we first study the number of packets that the origin router and the destination router receive, which are denoted as n_1 and n_2 respectively. We randomly pick 100 OD pair in the traffic matrices and present the values of n_1 and n_2 in Fig. 4.6. The x-axis is the index of the OD pair. Each index corresponds to an Origin–Destination pair, (n_1, n_2). The figure shows that the values of n_1 and n_2 are very different from each other in most cases. For example, for the tenth OD pair, $n_1 = 1,111,022$ and $n_2 = 17,795,961$. The ratio between n_1 and n_2 is about 0.06. As the previous section shows, ODFM is not able to work well in this situation. We will further demonstrate that shortly.

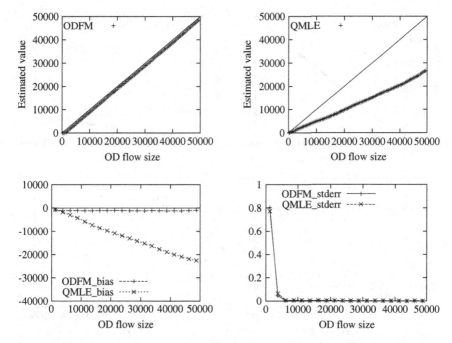

Fig. 4.3 *First plot* (*top left*) estimation results by ODFM when $n_1 = 6,000,000$ and $n_2 = 300,000$. *Second plot* (*top right*) estimation results by QMLE when $n_1 = 6,000,000$ and $n_2 = 300,000$. *Third plot* (*bottom left*) bias of ODFM and QMLE, which is the measured $E(\hat{n}_c - n_c)$ with respect to n_c. *Fourth plot* (*bottom right*) standard deviation of ODFM and QMLE, which is the measured $\frac{\sqrt{Var(\hat{n}_c)}}{n_c}$

Table 4.5 Number of memory accesses and number of hash operations per packet

	memory accesses	hash operations	constant?
ODFM	1	1	Yes
QMLE	1.17	2	No

4.4.2 Processing Overhead

Table 4.1 shows the averaged results of the per-packet processing overhead in terms of the number of memory accesses and the number of hash operations for each packet. ODFM requires only 1 hash operation and 1 memory access (memory write) for each packet. QMLE requires more per-packet processing overhead than ODFM. It incurs 1.17 memory accesses for each packet and 2 hash operations on average. Furthermore, ODFM requires constant per-packet processing overhead. While QMLE requires unpredicted per-packet processing overhead in terms of memory accesses.

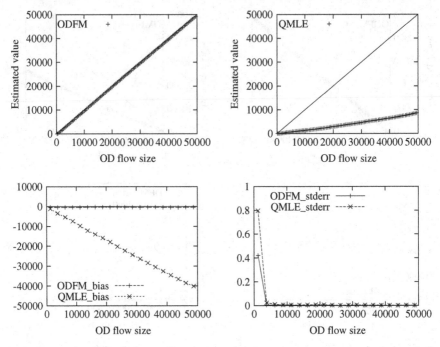

Fig. 4.4   *First plot (top left)* estimation results by ODFM when $n_1 = 6{,}000{,}000$ and $n_2 = 100{,}000$. *Second plot (top right)* estimation results by QMLE when $n_1 = 6{,}000{,}000$ and $n_2 = 100{,}000$. *Third plot (bottom left)* bias of ODFM and QMLE, which is the measured $E(\hat{n}_c - n_c)$ with respect to n_c. *Fourth plot (bottom right)* standard deviation of ODFM and QMLE, which is the measured $\dfrac{\sqrt{Var(\hat{n}_c)}}{n_c}$

4.4.3 Measurement Accuracy

Figure 4.7 has four plots. The first plot presents the estimation results of ODFM. The second plots presents the estimation results of QMLE. The third plot shows the corresponding estimation bias and the last plot shows the standard deviation. Clearly ODFM works far better than QMLE, which agrees with the simulation results in Fig. 4.5. The reason is that the origin router and the destination router are likely to receive different numbers of packets. And the performance of QMLE will degrade in that situation, while ODFM does not have this problem.

4.5 Other Methods

The origin–destination (OD) flow measurement methods mainly fall into two categories. One is *intermediate based* [14, 15, 17, 20, 23–25] and the other is *end-to-end based* [4, 9, 10]. The intermediate based methods [23–25] employ statistical

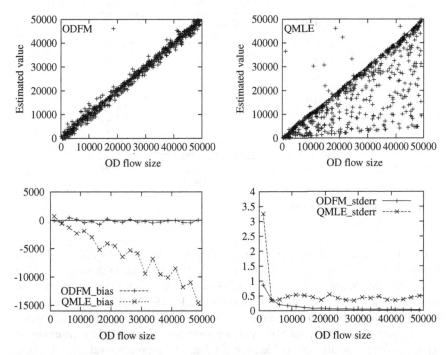

Fig. 4.5 *First plot* (*top left*) estimation results by ODFM when the values of n_1 and n_2 are randomly assigned between 100,000 and 10,000,000. *Second plot* (*top right*) estimation results by QMLE when the values of n_1 and n_2 are randomly assigned between 100,000 and 10,000,000. *Third plot* (*bottom left*) bias of ODFM and QMLE, which is the measured $E(\hat{n}_c - n_c)$ with respect to n_c. *Fourth plot* (*bottom right*) standard deviation of ODFM and QMLE, which is the measured $\frac{\sqrt{Var(\hat{n}_c)}}{n_c}$

techniques to indirectly estimate the OD flows based on link load, network routing, and configuration data, which are widely available information. Zhang et al. [23] assume an underlying gravity model [15, 19] for OD flows and use edge link load data together with additional information on intermediate routers to analyze the model. After that, they introduce the tomographic method [5, 7] to determine the results that most fit with the obtained gravity model. The methods in [24, 25] extend the point-to-point measurement to point-to-multipoint measurement using a regularization based on entropy penalization. These intermediate-based methods share a common property that jeopardizes them from being widely applied: the estimation relies on traffic volumes, which are usually unknown information. As a result, these methods either cannot achieve high measurement accuracy or incurs severe computational cost.

Considine et al. [8] use the method of moments for OD packet counts, which extracts a traffic digest from the packet stream. As the study in [4] points out, when the noise-to-signal ratios are high, the performance of [8] will be degraded.

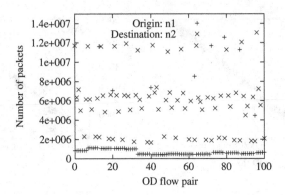

Fig. 4.6 Number of packets for 100 OD pair

Cao, Chen and Bu [4] design a quasi-likelihood approach (QMLE) for OD flow measurement based on a continuous variant of the Flajolet–Martin sketches [11]. The approach maintains an array of buckets, whose initial values are all set to infinity at the beginning of a measurement period, in each network node. When it receives a packet, the node performs two hash operations. The first one pseudo-randomly chooses a bucket i in the array for packet information storage. The second one generates an exponential random number v based on the packet, whose expected value is one. After the two hash operations, the node updates the bucket i by v. If the original value of i is larger than v, the node will set the value of bucket i to v. Otherwise, it will skip this packet. At the end of the measurement period, in order to estimate the OD flow size of two routers r_1 and r_2, QMLE derives the quasi-probability distribution of the packet information and employs the maximum likelihood estimation to compute the OD flow size based on the values of the two bucket arrays.

QMLE is able to achieve small per-packet update overhead and accurate measurement result with a compact memory requirement. However, for each packet, it needs to perform two hash operations and more than one memory access on average, while the optimal should be exactly one hash operation and one memory access per-packet. Moreover, it also has to improve in terms of measurement accuracy, as is demonstrated by simulations and experiments in Sects. 4.3 and 4.4, respectively.

Also related is to recover the missing values during traffic measurement by the technique of compressive sensing in [26], which proposes a spatio-temporal framework to exploit the presence of both global structure and local structure. Rincon et al. [18] provide a multi-resolution analysis to develop a general model for traffic matrices, which is based on the diffusion wavelet transform. They find that the model must be sparse and also demonstrate it by experimental results.

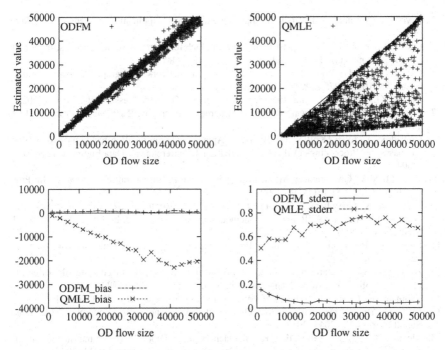

Fig. 4.7   *First plot (top left)* estimation results by ODFM when $n_1 = 1,000,000$ and $n_2 = 1,000,000$. *Second plot (top right)* estimation results by QMLE when $n_1 = 1,000,000$ and $n_2 = 1,000,000$. *Third plot (bottom left)* bias of ODFM and QMLE, which is the measured $E(\hat{n}_c - n_c)$ with respect to n_c. *Fourth plot (bottom right)* standard deviation of ODFM and QMLE, which is the measured. $\frac{\sqrt{Var(\hat{n}_c)}}{n_c}$

4.6 Summary

This chapter presents a novel method for OD flow measurement that employs the bitmap data structure for packet information storage and uses statistical inference to extract information from bitmaps. The method not only requires smaller per-packet processing overhead but also achieves much better accurate results, when comparing with the existing approach of QMLE. We use both simulations and experiments to demonstrate the superior performance of this method.

References

1. Abilene   Update.   http://www.internet2.edu/presentations/spring03/20030410-Abilene-Corbato.pdf (2003)
2. Abilene Network. http://en.wikipedia.org/wiki/Abilene_Network (2011)

3. Bryc, W.: The normal distribution: characterizations with applications. Springer,New York (1995)
4. Cao, J., Chen, A., Bu, T.: A quasi-likelihood approach for accurate traffic matrix estimation in a high speed network. In: Proceedings of IEEE INFOCOM (2008)
5. Cao, J., Davis, D., Wiel, S.V., Yu, B.: Time-varying network tomography. J. Amer. Statist. Assoc. **95**, 1063–1075 (2000)
6. Casella, G., Berger, R.L.: Statistical Inference, 2nd ed. Duxbury Press, Pacific Grove (2002)
7. Coates, M., Hero, A., Nowak, R., Yu, B.: Internet tomography. IEEE Signal Process. Mag. 19(3), 47–65 (2002)
8. Considine, J., Li, F., Kollios, G., Byers, J.: Approximate aggregation techniques for sensor databases. In: Proceedings of the 20th International Conference on Data Engineering (ICDE) (2004)
9. Duffield, N.G., Grossglauser, M.: Trajectory sampling for direct traffic observation. In: Proceedings of ACM SIGCOMM (2000)
10. Feldmann, A., Greenberg, A.G., Lund, C., Reingold, N., Rexford, J., True, F.: Deriving traffic demands for operational IP networks: methodology and experience. In: Proceedings of ACM SIGCOMM (2000)
11. Flajolet, G.: Probabilistic counting. In: Proceedings of Symposium on Fundations of Computer Science (FOCS) (1983)
12. Hwang, K., Vander-Zanden, B., Taylor, H.: A linear-time probabilistic counting algorithm for database applications. ACM Trans. Database Syst. **15**(2), 208–229 (1990)
13. Lehmann, E., Casella, G.: Theory of Point Estimation. Springer Press, New York (1998)
14. Liang, G., Yu, B.: Maximum pseudo likelihood estimation in network tomography. IEEE Trans. Signal Process. **51**, 2043–2053 (2003)
15. Medina, A., Taft, N., Salamatian, K., Bhattacharyya, S., Diot, C.: Traffic matrix estimation: existing techniques and new directions. In: Proceedings of ACM SIGCOMM (2002)
16. Newey, W., McFadden, D.: Large sample estimation and hypothesis testing. Dan. Handb. Econom. **4**, 2111–2245 (1994)
17. Nucci, A., Cruz, R., Taft, N., Diot, C.: Design of igp link weight changes for estimation of traffic matrices. In: Proceedings of IEEE INFOCOM (2004)
18. Rincon, D., Roughan, M., Willinger, W.: Towards a meaningful MRA of traffic matrices. In: Proceedings of ACM SIGCOMM IMC (2008)
19. Roughan, M., Greenberg, A., Kalmanek, C., Rumsewicz, M., Yates, J., Zhang, Y.: Experience in measuring backbone traffic variability: models, metrics, measurements and meaning. In: Proceedings of ACM SIGCOMM Internet Measurement, Workshop (2002)
20. Soule, A., Nucci, A., Cruz, R., Leonardi, E., Taft, N.: How to identify and estimate the largest traffic matrix elements in a dynamic environment. In: Proceedings of ACM Sigmetrics (2004)
21. National Institute of Standard and Technology: FIPS 180–1: Secure Hash Standard. http://csrc.nist.gov (1995)
22. Zhang, Y.: 6 months of Abilene traffic matrices. http://www.cs.utexas.edu/yzhang/research/AbileneTM/ (2004)
23. Zhang, Y., Roughan, M., Duffield, N., Greenberg, A.: Fast accurate computation of large-scale ip traffic matrices from link loads. In: Proceedings of ACM SIGMETRICS (2003)
24. Zhang, Y., Roughan, M., Lund, C., Donoho, D.: An informationtheoretic approach to traffic matrix estimation. In: Proceedings of ACM SIGCOMM (2003)
25. Zhang, Y., Roughan, M., Lund, C., Donoho, D.: Estimating point-to-point and point-to-multipoint traffic matrices: an information-theoretic approach. IEEE/ACM Trans. netw. **13**(5), 947–960 (2005)
26. Zhang, Y., Roughan, M., Willinger, W., Qiu, L.: Spatio-temporal compressive sensing and internet traffic matrices. In: Proceedings of ACM SIGCOMM (2009)